Strategic Issues in Public–Private Partnerships

An international perspective

Mirjam Bult-Spiering

Geert Dewulf

Blackwell
Publishing

© 2006 Mirjam Bult-Spiering and Geert Dewulf

Blackwell Publishing Ltd editorial offices:
Blackwell Publishing Ltd, 9600 Garsington Road, Oxford OX4 2DQ, UK
   Tel: +44 (0)1865 776868
Blackwell Publishing Inc., 350 Main Street, Malden, MA 02148-5020, USA
   Tel: +1 781 388 8250
Blackwell Publishing Asia Pty Ltd, 550 Swanston Street, Carlton, Victoria 3053, Australia
   Tel: +61 (0)3 8359 1011

First published 2006 by Blackwell Publishing Ltd

ISBN-10: 1-4051-3475-5
ISBN-13: 978-1-4051-3475-0

Library of Congress Cataloging-in-Publication Data

Bult-Spiering, Mirjam.
    Strategic issues in public–private partnerships : an international perspective / Mirjam Bult-Spiering, Geert Dewulf.
     p. cm.
    Includes bibliographical references and index.
    ISBN-13: 978-1-4051-3475-0 (hardback: alk. paper)
    ISBN-10: 1-4051-3475-5 (hardback: alk. paper)   1. Public–private sector co-operation.
    2. Public–private sector co-operation–Europe.   3. Public–private sector co-operation–United States.   I. Dewulf, Geert.   II. Title.
HD2961. B79 2006
303. 3—dc22
2005034200

A catalogue record for this title is available from the British Library

Set in 10/12.5 pt Minion
by Graphicraft Limited, Hong Kong
Printed and bound in Singapore
by COS Printers Pte Ltd

The publisher's policy is to use permanent paper from mills that operate a sustainable forestry policy, and which has been manufactured from pulp processed using acid-free and elementary chlorine-free practices. Furthermore, the publisher ensures that the text paper and cover board used have met acceptable environmental accreditation standards.

For further information visit our website:
www.blackwellpublishing.com/construction

Contents

About the authors

Mirjam Bult-Spiering is Assistant Professor of Public–Private Governance, Department of Construction Management and Engineering at the University of Twente, the Netherlands. w.d.bult-spiering@ctw.utwente.nl

Geert Dewulf is Professor of Planning and Development, Department of Construction Management and Engineering at the University of Twente, the Netherlands. g.p.m.r.dewulf@ctw.utwente.nl

Contributors

Anneloes Blanken is a PhD student in the Department of Construction Management and Engineering at the University of Twente, the Netherlands. a.blanken@ctw.utwente.nl

Gerrit-Jan Knaap is Professor of Urban Studies and Planning and Director of the National Center for Smart Growth Research and Education at the University of Maryland, USA. gknaap@umd.edu

Marnix Smit is a PhD student in the Department of Construction Management and Engineering at the University of Twente, the Netherlands. m.smit@ctw.utwente.nl

Preface

'No profit grows where is no pleasure ta'en; In brief, sir, study what you most affect' (Tranio in Shakespeare's *The Taming of the Shrew*, Act 1 Sc. 1, 39).

Shakespeare did not write a book on public–private partnerships (PPPs) but this statement shows great wisdom valuable for all stakeholders involved in PPPs. Discussions on PPPs have focused primarily on profits and not on the value for society or less quantifiable long term benefits. Value for money is, however, the key driver for governments to launch PPPs. Value for money is a concept that is politically easy to sell but hard to quantify.

The idea of writing a book on strategic issues in PPPs started years ago when we organized workshops with senior civil servants and captains of industry to discuss the future of PPPs. What struck us was the political rather than rational character of the debate, as well as the focus on short term aims and money. As we delved further into the subject of PPPs, we increasingly found that actions of politicians, private parties and interest groups were governed by subjective judgements and not by insight. Moreover, the term 'PPP' was being confused with outsourcing and privatization, which impeded open debate. With this book, we intend to stimulate a more strategic debate on PPPs. The decision to start a PPP should be a strategic one, dependent on the goals and values governments and business wish to achieve. 'Study what you most affect!'

This book provides a theoretical foundation for the analysis of the creation and functioning of PPPs, illustrated by several examples, especially from Europe and the United States. It is based on a thorough review of the literature on PPPs as well as several research projects we have undertaken in recent years. It was our intention to give the reader a broad picture of PPP developments world-wide, an insight into the various forms of PPP, and information about the motives and rationale behind PPPs, to enable more strategic decisions to be made on PPPs. Using the research findings, starting points for the enhancement of PPPs are presented throughout the book.

In Chapter 2, a framework for the description of PPPs is presented, and the chapter ends with an overview of problems in PPP practice. Chapter 3 covers in broad terms some characteristics of procurement in the construction sector, as well as the features of two dominant PPP procurement systems: concessions and joint ventures. Chapter 4 and Chapter 6 each explain both types of PPP in detail, with Chapter 4 focusing on concession PPPs and Chapter 6 on joint venture PPPs. Illustrations of both types of PPP are presented in Chapters 5 and 7 respectively. Chapter 8 translates the findings in the previous chapters to current trends in PPPs and outlines future (research) assignments for and developments in PPPs.

Public–private partnership is taught as a concept of modern governance at universities and in various courses for practitioners. The purpose of this book is to provide researchers with a better understanding of the various PPP concepts. The book is also aimed at developing more capable government and private management of PPP projects. The book is not written as a manual, since managing a PPP is not a routine job. Every PPP project is different and the way the process should be managed will differ in each case. Managing PPPs requires considerable capabilities and skills from both public and private managers, so the book aims to deliver insight and tools for those managers. It also seeks to launch a strategic debate and to develop a research agenda for the future. There is still a large gap in our knowledge on PPPs and much work remains to be done. This book fills only a small part of the gap.

We hope that we have succeeded in our goals and that we have delivered value for money to our readers.

Geert Dewulf
Mirjam Bult-Spiering
Enschede

Acknowledgements

We are indebted to many people and institutions. During our research we had the chance to discuss the topic of public–private partnership (PPP) with many colleagues, practitioners, students and politicians. Several workshops and symposia with practitioners have delivered input and material for this book. The authors also supervised many graduates and PhD students working on PPP related themes, most of whom wrote their theses while working in practice, which delivered not only interesting academic insights but also information about interesting cases. Working with students and practitioners not only provided information and insight, it helped us formulate the need for further knowledge on PPPs and set up a demand-oriented research agenda.

Anneloes Blanken and Marnix Smit, both working on their PhD theses, read and discussed with us all the chapters at different stages in the process; Marnix is co-author of Chapter 6 and Anneloes of Chapter 8. Their contributions improved the quality of the book enormously. We further profited from interesting discussions on individual chapters with our departmental colleague Andre Dorée and Gerrit Knaap from the University of Maryland. Gerrit Knaap is also the co-author of Chapter 7. We should like to thank all our colleagues in the Department of Construction Management and Engineering at the University of Twente for the numerous discussions we had on the topic of PPP.

This book would not have been possible without the sponsors of the many research projects we have undertaken in recent years. An important base for this book was formed by the international study on PPP we undertook in 2003–2004. This study was sponsored by a large number of private and public parties: Ballast Nedam; Dura Vermeer Group; Grontmij Group; Heijmans; Royal BAM Group; the Dutch Ministry of Spatial Planning, Housing and Environment; the Ministry of Transport, Public Works and Water Management; PPP Knowledge Centre; ProRail; Royal Haskoning; Strukton Group; TBI; and Volker Wessels. They not only sponsored our research, but also provided information about projects and gave us interesting contacts in several countries. For this book a large number of institutions were interviewed and provided us with all kinds of documentation. We should like to thank Gerardo Gavilanes Gineres of the Spanish Ministerio de Fomento and Rui Sousa Monteiro of the Portuguese Parpublica for introducing us to, respectively, Spanish and Portuguese stakeholders as well as the many other people who helped us during our research.

Special thanks go to the people who gave their time to discuss PPPs with Mirjam during her research for this book in the United States: Gerrit Knaap and his National Center for Smart Growth Research and Education hosted her sabbatical leave; John Donahue (Harvard University), Scott Fosler, Bill Lucyshyn and Kim Ross (University of Maryland), Brad Gentry (Yale University), Julia Rubin (Rutgers University), Gregory Stankiewicz (Princeton University) and Lawrence Susskind (Massachusetts Institute of Technology) were kind and enthusiastic enough to share their knowledge on the subject.

Julia Burden and Emma Moss of Blackwell Publishing have been unstinting in their patience and support. We hope that the publisher is pleased with the result.

Our partners and children have suffered under the burden of our working on this book. We can now pay them the attention they deserve.

1 Introduction

Many governments throughout the world have been involved in public–private partnerships (PPPs). Shifts in the boundaries between the public and private sectors caused by several movements explain this trend.

Governments increasingly depend on the private sector for the implementation of projects. The public sector can no longer afford large investments, so private sector involvement is required. This changes the relationship between government and marketplace. Boundaries between public and private organizations are blurring and public management is changing. Output and performance are now the key criteria for evaluating the public sector. The attention on PPPs fits this context: public and private sectors combine their efforts.

Public–private partnerships are a special feature of governance. To understand PPPs, we need to understand governance. A focus on governance means a focus on process, rather than on institutions. Governance comprises institutions of government and the processes through which these institutions interact with civil society (Pierre, 1997a). Self-organizing, complex and dynamic inter-organizational networks are characteristics of today's social political world (Stoker, 1997; Laws et al., 2001). The term 'governance' recognizes the interdependence of organizations and tries to meet the requirements of this world (Stoker, 1997).

Governance is a study not only of organizations but also of the public sector as a whole, and consequently the relations between all players in the public domain: local, regional and central governments, private parties, citizens and interest groups. Good governance therefore implies the involvement and endeavour of all players (Montfort, 2004).

Pierre & Peters (2000) list eight factors that focused increasing attention on governance issues in the last decade of the twentieth century:

(1) A shift from politics toward the market
(2) Economic crises forced governments to consider their management intentions in perspective
(3) Economic and political globalization forced adjustment of management intentions and changes in public institutions at a national level
(4) Dissatisfaction with the government's performance led to more private sector involvement in the enforcement of public tasks
(5) The rise of 'new public management' (NPM) ideas and instruments as a counterpart to traditional, input-oriented management styles resulted in an increasing interest in output management, evaluations and separation of policy making and enforcement

(6)   Increasing attention on citizen participation and environmental issues

(7)   The increasing importance of sub-national and super-national institutions has resulted in a growing interest in multi-level governance

(8)   The tension between several new forms of public management and co-ordination on the one hand, and the old traditions and instruments of public accountability on the other, resulted in the introduction of 'new' players (private parties, citizens, non-governmental organizations (NGOs)) into the political process.

The central issue in these explanations is the changing role of government:

'We believe that the role of the state is not decreasing as we head into the third millennium but rather that its role is transforming, from a role based in constitutional powers towards a role based in coordination and fusion of public and private resources.' (Pierre et al., 2000)

The focus on governance rather than on government emphasizes the increasing participation of multiple parties with certain stakes. This is true of all activities in the public sector domain, including spatial planning and construction projects.

The debate on governance has three levels of focus (Montfort, 2004):

(1)   The individual organizations

(2)   The public sector as a whole

(3)   Interactions and links between parties in the public domain.

The third level debate is most pertinent to PPPs. Governance is therefore interpreted in this book as the functioning of networks and co-ordination mechanisms. In Chapter 2, the consequences of this approach are further specified.

This book analyses and describes the various forms of PPP. The book cannot be seen as a manual for managing PPP projects since every project is different and tailor-made solutions are needed. Nevertheless, the book contains many lessons for public and private managers as well as for researchers and students in the field of PPPs. Most PPP research focuses primarily on contracts, risks and financial or legal arrangements. Numerous studies are available on procurement, contracts and financial structures of PPPs. However, little is known about the way public and private representatives interact during the formation and operation of the partnerships. Several evaluations have shown that the success of PPPs strongly depends on the way this process is managed, or the way interests and interactions between partners are managed (Pierre, 1997b; Spiering & Dewulf, 2001; Fischbacher & Beaumont, 2003). Much attention is paid to the process aspects and dynamics of PPPs in construction in this book.

1.1 Public–private spectrum

The central question on governance from the perspective of PPPs is how to organize the interaction between public and private sector. The main goal is to

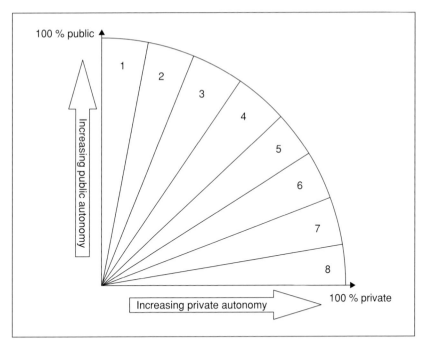

100 % public

Increasing public autonomy

1
2
3
4
5
6
7
8

Increasing private autonomy

100 % private

Figure 1.1 The public–private spectrum (Börzel & Risse, 2002). Reproduced with permission of Freie Universitat, Berlin.

improve efficiency, quality of public services and products, and legitimacy. The question how to organize a PPP cannot be answered in general: for every market, and in most cases even for every project, the answer has to be tailor-made.

Confusion about the PPP concept is striking in the political and social discussion on these governance questions. Often, PPP is used as a synonym for privatization. Nevertheless, there are significant differences between PPP and privatization. In PPPs, public and private parties (actors) *share* costs, revenues and responsibilities. Privatization represents the *transfer* of tasks and responsibilities to the private sector, with both costs and revenues being in private hands. The confusion impedes a rational discussion about PPPs since all the disadvantages of privatization are imputed to PPPs.

The difference between PPPs and privatization can be visualized in a public–private spectrum: PPP is an organizational structure somewhere in the middle between public and private regimes. Börzel & Risse (2002) picture this spectrum as given in Fig. 1.1. At one end of the spectrum, the authors define public regulation with no involvement of private parties, and at the other private self-regulation with no public involvement. In between, they distinguish lobbying, consultation, co-regulation and delegation. In short, Börzel & Risse (2002) define the following forms of organization:

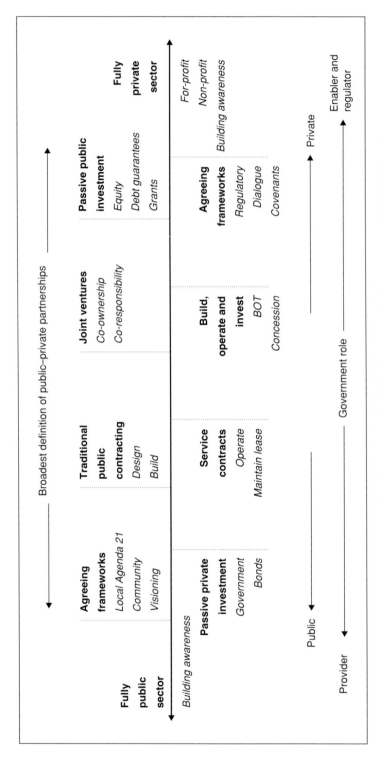

Figure 1.2   Spectrum of public–private partnerships (Bennett et al., 2000). Reproduced with permission of PPPUE, New York.

(1) Public regulation: no involvement of private parties
(2) Lobbying of public parties by private parties
(3) Consultation and co-option of private parties
(4) Co-regulation of public and private parties (e.g. private parties as negotiation partners): joint decision making by public and private parties
(5) Delegation to private parties (e.g. standard setting): participation of public parties
(6) Private self-regulation in the shadow of hierarchy (e.g. voluntary agreements): involvement of public parties
(7) Public adoption of private regulation: output control by public parties
(8) Private self-regulation (purely private regimes): no public involvement.

Bennett et al. (2000) use a broader definition of PPPs than co-regulation (list point 4), in which forms (2) to (6) represent PPP. The different PPP alternatives they consider possible within these phases are presented in Fig. 1.2.

Bennett et al. (2000) consider the term public–private partnership to be a spectrum of possible relationships between public and private parties for the co-operative provision of infrastructure services (Fig. 1.2). Savas (2000), for example, uses a similar spectrum ranging from fully public (which he calls 'government department') to fully private ('build–own–operate'). The various alternatives described by Bennett et al. (2000) within this spectrum are as follows:

■ *Awareness building* Building awareness of either governments or private parties of the opportunities for improving the delivery of services through the spectrum of collaborative approaches
■ *Agreeing frameworks* Agreeing on the basic frameworks for community or private action through participatory mechanisms
■ *Passive private investment* Making private investment available to government-run operations
■ *Traditional public contracting* Entering into a contractual relationship under which the public sector purchases a product from the private sector or hires a private organization to design or build a new facility
■ *Operation, maintenance and service contracts* Entering into a contractual relationship under which the public sector essentially hires a private organization to carry out one or more specified tasks or services for a specified period. The public sector remains the primary provider of the service and is responsible for funding any capital investments needed to expand or improve the system
■ *Joint ventures (mixed-capital partnerships)* Creating a company jointly owned by the government and private companies, in which they assume co-responsibility for the delivery of services. The public and private sector partners can either hold shares in a new company or assume joint ownership of an existing company which provides services. Joint ventures require that

both parties accept the idea of shared risk and shared reward. Each must be willing to make quantifiable contributions throughout the project development and implementation process

■ *Build–operate–transfer (BOT) contracts* Bringing private investment into the construction of new facilities, or the substantial renovation of existing ones by using BOT contracts. Under a BOT contract, the private sector finances, builds and operates a facility or system according to performance standards set by the government for time periods ranging from 10 to 20 years. The government retains ownership of the facility and becomes both the customer and the regulator of the service

■ *Concessions* Awarding a private firm (concessionaire) full responsibility for the delivery of services in a specified area, including all related operational, maintenance, fee collection and management activities. The concessionaire is also responsible for any capital investments required to build, upgrade or expand the system, as well as for financing those investments using the tariffs paid by system users. The government is responsible for establishing performance standards and ensuring that the concessionaire meets them. In essence, the public sector's role shifts from being the provider of the service to being the regulator of its price and quantity. The fixed infrastructure assets are entrusted to the concessionaire for the duration of the contract (typically about 25–30 years), but they remain government property

■ *Passive public investment* Bringing government funds to private operations through the grants, equity investments, loans or guarantees offered by a range of public institutions

■ *Community-based provision* Using any of the options listed above or others to help communities address their own needs. Community-based provision starts when financial or institutional limitations prevent the government from providing adequate services to particular sectors of the population, forcing residents to find their own means of meeting their needs. Community-based providers might include individuals, families or local micro-enterprises. Community-based organizations often play a key role in organizing poor residents into taking collective action and in representing their interests in negotiations with NGOs and governments. Community-based provision schemes may also be integrated with the formal systems run by the public, the formal business sector or both.

The only essential ingredient in Bennett's model is some degree of private participation in the delivery of traditionally public-domain services. There are almost infinite numbers of individual points on the spectrum, with new models for co-operation being developed all the time. The broadest definition of a PPP is any type of voluntary co-operation between public and private parties (Bennett et al., 2000).

In this book, we use a narrower definition of PPP, confining ourselves to *concession* PPPs (including BOT contracts) and *joint-venture* PPPs. Concessions are described extensively in Chapters 4 and 5, and joint ventures in Chapters 6 and 7. In general, concession PPPs in construction are found for infrastructure development in a broad sense and therefore operate at local, regional or national government level, whereas joint venture PPPs are applied to urban development projects and function merely at local government level. The contractual forms of concession PPPs that we distinguish are build–operate–transfer (BOT) and design–build–finance–maintain/operate) (DBFM/O). In joint venture PPPs, public and private partners are shareholders of a joint public–private company.

In the construction sector, the term PPP is often bracketed with organizational structures such as alliances and partnering. Alliances provide non-adversarial incentive arrangements for project coalition members on a single project (Winch, 2002). In alliance contracts, most risks are allocated to either the public or the private party. Risks that can be influenced by both the public and the private parties are set apart and the risk premiums are deposited in an alliance fund; these risks and the fund are managed collectively (Knibbe & Spiering, 2003). In contrast with PPPs, not all risks are part of the public–private agreement.

Partnering refers to the formal arrangements between members of a project coalition under which they work together on a programme of projects (Winch, 2002). Partnering mostly concerns private–private co-operation and is based on the principles of supply chain management: private partners make production arrangements for a programme of projects that have a fixed chain of suppliers.

In this book we make a clear distinction between PPP and alliances and partnering: alliances and partnering therefore will not be elaborated any further.

1.2 History

Public–private partnerships have a long history in many countries, but grew significantly more popular during the 1980s. At this point, private sector thinking was introduced and used in the public sector, and market-based criteria were applied to the delivery of public products and services (Pierre, 1997a). During the 1990s, NPM and market-based philosophies further influenced public management in many countries. Because the degree of complexity of the problems needing to be solved increased as a result of growing interdependencies between assignments and parties involved, more partnerships between public and private sectors were formed.

Public–private partnerships have the longest tradition in the USA. In the 1950s and 1960s, PPPs in the USA were set out by the federal government as a tool for stimulating private investment in inner-city infrastructure and regional economic development (Fosler & Berger, 1982; Beauregard, 1997; Linder, 1999). They became an explicit instrument during President Carter's administration: the 1978 national urban policy and urban development action grant (UDAG) encouraged cities to go from private investment subsidies to joint equity venture PPPs

(Stephenson, 1991; Clarke, 1998; Linder, 1999; Rubin & Stankiewicz, 2001). The Reagan administration reinforced this orientation towards private investment by reducing the size of federal government and its role in local decision making; policies gave priority to private investment decisions (Clarke, 1998; Walzer & York, 1998).

Throughout the 1980s, PPPs increasingly became a derivative of the privatization movement and government rethinks. Private providers were assumed able to provide higher quality goods and services at lower cost, thereby significantly reducing the government's tasks and responsibilities (Linder, 1999). Hence, the factors that explained the emergence of PPPs in the Carter and Reagan administrations were the decreased role of national government, a declining faith in government, and the need for private capital and its effective use (Beauregard, 1997).

The Clinton administration, advocate of the Third Way policies, promoted PPPs as a key component of its urban policy. Third Way policy ideas are consistent with NPM, viewing partnerships as a new way of governance (Rubin & Stankiewicz, 2001). PPPs are considered to be NPM mechanisms and a means of establishing new forms of governance working across organizational boundaries (Keating, 1997). Unlike earlier federal governments' partnership encouragement, the Clinton administration emphasized the importance of full local community involvement (Rubin & Stankiewicz, 2001). The dependence of local government on business investment is strong in the USA, because of the absence of state and federal aid.

It was not only in the USA that PPPs assumed greater importance in the latter half of the twentieth century. In Spain, early examples occurred in the 1960s and toll roads had already been developed by 1968.

In the UK, the 1979 Conservative government believed that central government was too involved in the economy and needed to step down in favour of utilizing private capital. Enterprise zones and urban development corporations (UDCs) (business interest bodies given substantial government funding and responsibility for economic development in declining areas) were important instruments in this ambition.

In the UK in the late 1980s, the Thatcher administration turned to PPPs as the preferred method for economic regeneration. City Challenge (the programme that encouraged local authorities to propose schemes for economic regeneration in partnership with local businesses) replaced UDCs. The UK thinking on partnerships was significantly influenced by best practices in the USA.

Other parts of Europe also started using PPPs in the late 1980s. In The Netherlands, for example, the PPP idea was clearly introduced in the government's policy statement of 1986:

> 'New structures of public and private co-operation are founded, including local government, local business and, if necessary, central government, aimed at raising investments in urban renewal.' (Lubbers, 1986)

Since then PPP has been the centre of attention in The Netherlands; for example, the Minister of Public Works recently reinforced her decision to involve the private sector earlier and more intensely in pursuit of her policies. In The Netherlands, the emergence of PPPs can also be explained as resulting from the reconsidered role of government and the need for efficient and large-scale approaches (Kouwenhoven, 1991; Lemstra, 1996).

Even in a prosperous country such as Norway, PPPs have been introduced in the past decade. For years it was argued that Norway, because of its oil revenues, had little or no incentive to embark on risk sharing with the private sector (Greve, 2003). However, now the Norwegian government is co-operating with private parties in several infrastructure developments.

Examples of PPPs in developed countries can also be found outside Europe and the USA. In Australia, for example, the introduction of public–private arrangements for the provision of infrastructure dates back to the early 1990s. The first projects focused on toll roads, hospitals, water and power; in the mid-1990s, the focus was on prisons, sea ports and sports stadiums; in the late 1990s, airports were added to this list, with defence, schools and courts attracting contracts from 2001 (Crump & Slee, 2005). The introduction of PPPs is a reaction to the large costs and inherent risks in terms of cost recovery involved in the construction of many large infrastructure projects in Australia. Alongside some circumstantial and environmental factors, aspects that give PPPs their current appeal in the Australian context include the potential for achieving cost efficiencies, early project delivery, achieving gains from innovation, transferring some project and finance risk to the private sector, and creating and accessing improved services for citizens (English & Guthrie, 2003).

In many countries worldwide we see similar trends in private sector involvement and PPP developments. At first sight the rationale behind public–private co-operation is similar: in all countries, governments are relying increasingly on private sector money and skills. However, upon analysing the different forms of PPP, we notice major differences in the motives and procurement rules between countries. In Section 1.3, which describes the various PPP arrangements in more detail, we will elaborate on these historical differences.

Several publications indicate that partnerships have been and will continue to be an important economic development strategy, although their efficiency may be disputable (Clarke, 1998; Walzer & York, 1998; Linder, 1999; Rosenau, 1999; DiGaetano & Strom, 2003).

1.3 Global perspective

The national context is important in understanding PPPs. A global discussion helps us understand both PPP and the influence of its different contexts. To learn about PPPs in different countries, the factors that determine public–private relations and that typify certain fields of partnership must be recognized. Various

Table 1.1   National contexts of public–private partnerships.

	USA	Europe
Causes	Financial crisis in the public sector Increased mobility of capital Increased complexity of government tasks Dominance of neo-liberal ideas	
Autonomy	Private UK	Public
Bureaucracy	Fragmented competitive	Unitary central
Public–public relation	Independent	Dependent
Public–private relation	Stable UK	Weak
Leadership	Strongly organized local business elite	No local business leadership

forms of PPP will be discussed with the objective of enabling strategic choices to be made on which form is needed in practice. Primarily, we use PPP developments in the USA and Europe for our analysis of the various manifestations of PPP; where relevant we have added examples from other continents. We acknowledge some differences within Europe and will describe the major differences between the USA and Europe, often using the UK as an intermediate case; Table 1.1 summarizes the similarities and differences between PPP contexts and is further explained below.

Common causes for movement towards partnerships found in both the USA and Europe are as follows (Keating, 1997; Grimshaw et al., 2002; Flinders, 2005):

■ Fiscal crisis in the public sector, and therefore a search for other sources of funding
■ Increased mobility of capital, which causes a power shift in the relationship between government and capital towards the private sector
■ Increased complexity of government tasks requiring an overlap between the public and private sector
■ Dominance of neoliberal ideas, such as NPM, and the reliance on market mechanisms and incentives.

In general terms, the key difference between the USA and Europe is the traditional autonomy of the private sector (property rights) in the former versus the traditional autonomy of the public sector in the latter. The USA and continental Europe represent the two extremes, with the UK as an intermediate case (see, for example, DiGaetano & Strom, 2003).

The US system is fragmented and pluralistic: there is no such thing as 'the government'; in Europe, however, the government is a policymaker, and certain political movements dispute the privileges of private capital. The relationships between government and capital are stable in the USA, but are relatively weak in Europe. Whereas central bureaucracy is large and has a strong role in Europe, in the USA bureaucracy is fragmented with units that compete with each other.

In a territorial sense, differences appear in the relationship between local and central government. American states have a high degree of political and functional independence of the federal government, and the local governments have a high degree of autonomy within the states; government in Europe is more unitary. As a financial consequence, local governments in the USA depend less on the federal government and more on private capital than in Europe where local government is much more financially dependent upon central government (Keating, 1993).

In contrast to most European cities, American cities set their own tax rates and appraisals, and therefore give important incentives to compete with other communities for firms and households. States determine the tasks for which local authorities are responsible, there being large differences between states. Local governments depend economically on private investment activities to generate the tax revenues to sustain their governmental functions. In addition, significant zoning and the existence of land control authorities are incentives for local governments to work with private investors in urban development (Clarke, 1998). In addition, the private sector plays a larger role in politics in the USA than in Europe, and the local business elite ('local business leadership') is more strongly organized.

In general, countries with strong public traditions seem to generate PPPs that are dominated by public parties. In countries with a weaker public sector tradition, the private sector will dominate the partnership. Partnerships in a global perspective can be seen as a continuum, ranging from public to private sector dominance, as shown in Fig. 1.3 (Pierre, 1997b; Savitch, 1997). The balance of a partnership is therefore typically public sector dominated in Sweden and France, private sector dominated in the USA, and publicly managed in the UK. Savitch (1997) outlines this global range of partnerships in relation to voluntary traditions as pictured in Fig. 1.4.

In France, the so-called 'Societé d'Economie Mixte' (SEM) and 'Etablissement Publique' (EPA) are used to stimulate development and build housing. SEMs have private shareholders, but the majority of shares are owned by the public

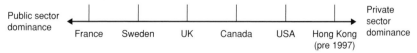

Figure 1.3 Public and private sector dominance. Reproduced with permission from Savitch, H.V. (1997) The ecology of public–private partnerships: Europe. In: J. Pierre (Ed) *Partnerships in Urban Governance: European and American Experiences* (pp. 175–186). London: Macmillan.

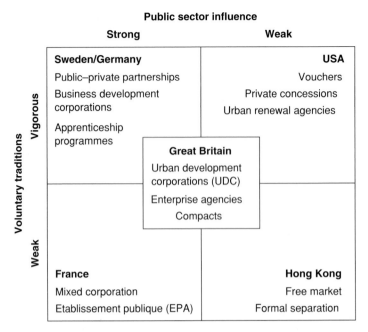

Figure 1.4   State, society and partnerships. Reproduced with permission from Savitch, H.V. (1997) The ecology of public–private partnerships: Europe. In: J. Pierre (Ed) *Partnerships in Urban Governance: European and American Experiences* (pp. 175–186). London: Macmillan.

sector and the public sector leads. EPAs take on contracts with private businesses and grant them concessions to operate, but again under complete public management. At the other end of the continuum, in the USA, the private provision of services is subsidized through public subsidies (vouchers) and the private sector has discretion in implementing public policies. The UK model is close to the European model, with a strong central government and powerful bureaucracy, but is closer to the American model in a cultural and ideological sense, the UK and the USA sharing similar values concerning the worth of private enterprise (Savitch, 1997; DiGaetano & Strom, 2003).

In a global historical perspective, we can state that PPPs in their current popular format first appeared in the USA as local, long term, policy-based joint venture partnerships. The UK adopted this model, just as several other European countries such as The Netherlands did, each with their own variations. The American examples of joint venture PPPs in the 1980s could, however, not be copied fully because of the weaker relationship between business and the public sector in Europe and the absence of local business leadership. Therefore, project-specific joint venture PPPs are dominant in Europe, whereas enduring policy-based partnerships dominated in the USA, at least in the 1960s and 1970s. (Although policy-based and institutionalized, many PPPs in the USA are project *driven* (Stephenson, 1991).)

Concession PPPs, which are project-based PPPs by definition, were used frequently in the USA before World War II, although not under the PPP heading. Afterwards, a stronger reliance on public means for infrastructure development occurred. In Europe, concession PPPs were used in Spain from the mid-1950s onwards, but came to prominence through the introduction of the UK's private finance initiative (PFI) policy. This successful policy was announced in the 1992 Autumn Statement and aimed for closer partnerships between the public and the private sector (House of Commons Treasury and Civil Service Committee, 1993). Later in the 1990s, these concession PPPs became popular all over Europe; their popularity is still growing.

1.4 Structure of this book

A great many misconceptions and confusion about the aims and contents of PPPs, as well as historical differences between the many countries where they are used, endanger well-considered employment of the PPP in realizing construction projects.

This book is dedicated to improving the performance of PPPs. Motives and starting points for the enhancement of PPPs are presented based on the findings of several research projects, literature reviews and case studies.

A framework for the description of PPPs is presented in Chapter 2; this chapter is based on an extensive literature search and empirical analysis from the PhD study of Mirjam Bult-Spiering. It ends with an overview of problems in PPP practice. Chapter 3 describes general characteristics of procurement in the construction sector, as well as features of the concession and joint venture PPP procurement systems. Chapters 4 and 6 each explain a type of PPP in detail: Chapter 4 is about concession PPPs and Chapter 6 about joint venture PPPs, both types being illustrated in Chapters 5 and 7, respectively. Chapter 5 presents some examples of infrastructure development in Europe; Chapter 7 describes two examples of urban development in the USA. Whereas Chapters 4 and 5 are based only on empirical research, Chapters 6 and 7 are founded on literature and document analysis results. Chapter 8 translates the findings in previous chapters to current shifts in PPPs, and outlines future (research) assignments for and developments in PPPs.

References

Beauregard, R.A. (1997) Public–private partnerships as historical chameleons: the case of the United States. In: J. Pierre (Ed) *Partnerships in Urban Governance: European and American Experiences* (pp. 52–70). London: Macmillan.

Bennett, E., James, S. & Grohmann, P. (2000) *Joint Venture Public–Private Partnerships for Urban Environmental Services.* New York: Public Private Partnerships for the Urban Environment (PPPUE).

Börzel, T.A. & Risse, T. (2002) Public–private partnerships: effective and legitimate tools of international governance? In: E. Grande & W. Pauly (Eds) *Complex Sovereignty: on the Reconstitution of Political Authority in the 21st Century* (unpublished).

Clarke, S.E. (1998) Economic development roles in American cities: a contextual analysis of shifting partnership arrangements. In: N. Walzer & B.D. Jacobs (Eds) *Public–Private Partnerships for Local Economic Development* (pp. 19–46). Westport: Praeger.

Crump, S.J. & Slee, R. (2005) Robbing public to pay private: two cases of refinancing education infrastructure in Australia. *Journal of Education Policy*, **20** (2), 249–264.

DiGaetano, A. & Strom, E. (2003) Comparative urban governance: an integrated approach. *Urban Affairs Review*, **38** (3), 356–395.

English, L.M. & Guthrie, J. (2003) Driving privately financed projects in Australia: what makes them tick? *Accounting, Auditing & Accountability Journal*, **16** (3), 392–511.

Fischbacher, M. & Beaumont, P.B. (2003) PFI, public–private partnerships and the neglected importance of process: stakeholders and the employment dimension. *Public Money & Management*, **23** (3), 171–177.

Flinders, M. (2005) The politics of public–private partnerships. *British Journal of Politics and International Relations*, **7** (2), 215–239.

Fosler, R.S. & Berger, R.A. (1982) *Public–Private Partnership in American Cities*. Lexington: Lexington Books.

Greve, C. (2003) Public–private partnerships in Scandinavia. *International Public Management Review*, **4** (2), 59–68.

Grimshaw, D., Vincent, S. & Willmott, H. (2002) Going privately: partnership and outsourcing in UK public services. *Public Administration*, **80** (3), 475–502.

House of Commons Treasury and Civil Service Committee (1993) *1992 Autumn Statement and the Conduct of Economic Policy*. London: HMSO.

Keating, M. (1993) The politics of economic development: political change and local development policies in the USA, Britain and France. *Urban Affairs Quarterly*, **28** (3), 373–396.

Keating, M. (1997) Commentary: public–private partnerships in the United States from a European perspective. In: J. Pierre (Ed) *Partnerships in Urban Governance: European and American Experiences* (pp. 163–174). London: Macmillan.

Knibbe, A. & Spiering, B. (2003) Roads, congestion and private involvement: the future. *Privatisation & Public Private Partnership Review 2003/2004*. Essex: Euromoney Yearbook.

Kouwenhoven, V.P. (1991) *Publiek–Private Samenwerking: Model of Model?* Delft: Eburon.

Laws, D., Susskind, L., Abrams, J., Anderson, J., Chapman, G., Rubenstein, E. & Vadgama, J. (2001) *Public Entrepreneurship Networks*. Cambridge, MA: Department of Urban Studies and Planning, Massachusetts Institute of Technology.

Lemstra, W. (1996) Samenwerking tussen overheid en bedrijfsleven: utopie of werkelijkheid? *Bedrijfskunde*, **68** (3), 44–50.

Linder, S.H. (1999) Coming to terms with the public–private partnership. *American Behavioral Scientist*, **43** (1), 35–51.

Lubbers, R.F.M. (1986) *Regeerakkoord Tweede Kabinet – Lubbers*. The Hague: Staatsuitgeverij (in Dutch).

Montfort, C. van (2004) *Ruimte voor Goed Bestuur: Tussen Prestatie, Process en Principe*. The Hague: Wetenschappelijke Raad voor het Regeringsbeleid.

Pierre, J. & Peters, G. (2000) *Governance, Politics and the State*. London: Macmillan.

Pierre, J. (1997a) Public–private partnerships and urban governance: introduction. In: J. Pierre (Ed) *Partnerships in Urban Governance: European and American Experiences* (pp. 1–10). London: Macmillan.

Pierre, J. (1997b) Conclusions. In: J. Pierre (Ed) *Partnerships in Urban Governance: European and American Experiences* (pp. 187–199). London: Macmillan.

Rosenau, P.V. (1999) The strengths and weaknesses of public–private policy partnerships: editor's introduction. *American Behavioral Scientist*, **43** (1), 10–34.

Rubin, J.S. & Stankiewicz, G.M. (2001) The Los Angeles community development bank: the possible pitfalls of public–private partnerships. *Journal of Urban Affairs*, **23** (2), 133–153.

Savas, E. (2000) *Privatization and Public–Private Partnerships*. New York: Chatham House.

Savitch, H.V. (1997) The ecology of public–private partnerships: Europe. In: J. Pierre (Ed) *Partnerships in Urban Governance: European and American Experiences* (pp. 175–186). London: Macmillan.

Spiering, W.D. & Dewulf, G.P.M.R. (2001) *Publiek–Private Samenwerking bij Infrastructurele en Stedelijke Projecten*. Enschede: P3BI.

Stephenson, M.O. (1991) Whither the public–private partnership. *Urban Affairs Quarterly*, **27** (1), 109–127.

Stoker, G. (1997) Public–private partnerships and urban governance. In: J. Pierre (Ed) *Partnerships in Urban Governance: European and American Experiences* (pp. 34–51). London: Macmillan.

Walzer, N. & York, L. (1998) Public–private partnerships in US cities. In: N. Walzer & B.D. Jacobs (Eds) *Public–Private Partnerships for Local Economic Development* (pp. 47–68). Westport: Praeger.

Winch, G.M. (2002) *Managing Construction Projects*. Oxford: Blackwell Science.

Further Reading

Bult-Spiering, M., Blanken, A. & Dewulf, G. (2005) *Handboek PPS*. Utrecht: Lemma.

Chandler, J.A. (1998) Regenerating South Yorkshire: how the public sector dominates business partnerships in Britain. In: N. Walzer & B.D. Jacobs (Eds) *Public–Private Partnerships for Local Economic Development* (pp. 157–176). Westport: Praeger.

Klijn, E.H. & Teisman, G.R. (2003) Institutional and strategic barriers to public–private partnership: an analysis of Dutch cases. *Public Money & Management*, **23** (3), 137–147.

Ministry of Public Works (2004) *Kaderbrief PPS en Innovatief Aanbesteden*. The Hague: Ministry of Public Works.

Peters, B.G. (1997) With a little help from our friends: public–private partnerships as institutions and instruments. In: J. Pierre (Ed) *Partnerships in Urban Governance: European and American Experiences* (pp. 11–33). London: Macmillan.

Pongsiri, N. (2002) Regulation and public–private partnerships. *International Journal of Public Sector Management*, **15** (6), 487–495.

2 Characteristics of Public–Private Partnerships

Public–private partnerships are recognized worldwide as an important means for undertaking several kinds of construction projects. PPPs are expected to generate added value for both the public and private sectors by solving complex problems. Although the benefits are obvious, it is not always easy to define PPP. The first section of this chapter provides a brief overview of definitions used in the literature. However, the definitions lack clarity, so further elaboration of PPPs is required in order to understand them. The second section of the chapter therefore distinguishes several general characteristics of the ideal model of PPP. In practice, of course, such an ideal type does not exist. The third section describes the dominant sectors of construction in which PPP is practised. The fourth and final section points out the main problems in PPP practice in relation to the ideal model characteristics.

2.1 Definitions

There is no straightforward definition of PPP. Many forms of public–private co-operation exist and numerous definitions are available to describe PPPs. Linder (1999) distinguishes six different uses of the term PPP, each having its own interpretation of what the partnerships and their purposes are.

(1) *PPP as management reform* An innovative tool that will change the way government functions. This is a mentoring kind of relationship, with flow of knowledge primarily from business to government
(2) *PPP as problem conversion* A fix for problems attending public service delivery. This yields the commercialization of problems to bait the marketplace
(3) *PPP as moral regeneration* Getting government managers involved as market participants. Partnerships affect the people involved
(4) *PPP as risk shifting* A means of responding to fiscal stringency on the part of the government and of getting private interests to sign on
(5) *PPP as restructuring public service* Adapting administrative procedures and coping with the demands from the sector's employees through partnerships
(6) *PPP as power sharing* Spread control horizontally. Business–government relations are altered through, (a) an ethos of co-operation and trust, (b) mutual beneficial sharing of responsibility, knowledge or risk, (c) give-and-take and negotiating differences.

Since these distinctions merely focus on the causes for starting PPPs, whereas in this chapter we are looking for characteristics of PPPs as formal arrangements, we need a more prescriptive definition.

Another characterization is described by Peters (1997):

(1) PPPs involve two or more actors (parties) at least one of which is a public entity
(2) Each of the participants is a principal and can bargain on its own behalf
(3) The partnership involves an enduring relationship among the actors, with some continuing interactions
(4) Each actor must be able to bring either material or non-material goods to the relationship
(5) All actors have a shared responsibility for the outcomes of their activities.

Many other definitions are, in fact, a variation on Peters' characterization. For example, Andersen (2004) defines four core elements in partnership arrangements:

(1) A partnership involves two or more actors, including governmental actors and business actors
(2) It is required that the participating partners are principals who are capable of acting on their own
(3) A continuity of relations is implied: this also includes relational contracts that are based on trustful negotiation and dialogue
(4) Actors involved must be willing to invest material and non-material re-sources in the partnership: separate organizational structures will be established to define objectives, tasks, financial platform and responsibilities.

Both Peters (1997) and Andersen (2004) stress the importance of sociological and economic aspects such as trust, interaction, willingness to invest and shared responsibility. They also point out that a PPP is a continuous process of interaction and negotiation. Besides, the set-up of a separate organization is considered important.

The importance of commitment and risk sharing is also stressed by Nijkamp et al. (2002) in their definition of PPP.

'A PPP is an institutionalised form of cooperation of public and private actors who, on the basis of their own indigenous objectives, work together towards a joint target, in which both parties accept investment risks on the basis of a predefined distribution of revenues and costs.'

In addition, Klijn & Teisman (2003) recognize these characteristics, but explicitly use the phrase 'added value'.

'PPPs can be defined as co-operation between public and private actors with a durable character in which actors develop mutual products and/or services and in which risk, costs, and benefits are shared. These are based on the idea of mutual added value.'

Other definitions focus on the gap that PPPs ought to fill, for example:

'A public–private partnership can be seen as an appropriate institutional means of dealing with particular sources of market failure by creating a perception of equity and mutual accountability in transactions between public and private organizations through co-operative behaviour.' (Pongsiri, 2002)

The characteristics described in the definitions above can be seen as prescriptions of how an ideal model of PPP should work to generate added value. The definitions already pinpoint some interesting conditions for success:

■ The aim for mutual added value and a joint goal
■ The actors' capability of negotiating on their own behalf
■ Revenues are shared in accordance with the actors' investments and risk acceptance
■ Formalized co-operation arrangements.

An important distinction is drawn between policy-based partnerships, which lay down a set of general rules for private investment and operation through co-operative ventures, and project-based partnerships, that focus on a specific site or circumstance (Stephenson, 1991; Dunn, 1999). Policy-based PPPs do not have a definable end-point, while project-based partnerships have a clear end. The project-based partnerships are more formal, project-specific efforts to mobilize public and private resources and share investment risks (Fosler & Berger, 1982; Davis, 1986; Stephenson 1991; Clarke, 1998). They aim to deliver a certain product, and can be characterized as offensive (Peters, 1997). In either case, the task is so complex and time-consuming that partnerships of a formal and stable nature are required (Beauregard, 1997). Also, from a transaction point of view, institutional characteristics are necessary to reduce transaction costs (Peters, 1997). Therefore, increasingly, project-specific partnerships are used rather than constant organizational forms.

In this book, PPPs are regarded as formal arrangements between public and private parties, and therefore the focus is on project-based partnerships. Within these project-based partnerships, a distinction is made between concession and joint venture PPPs. In the case of joint ventures, the public and private parties both bring in knowledge and resources. Concessions can best be defined as a form of outsourcing. The emphasis in joint venture PPPs is on togetherness or sharing, while concessions are about transfer of risks and revenues (Dewulf et al., 2004). Despite their long term focus, concession PPPs merely have the characteristics of project-based partnerships. Examples of joint venture PPPs in the different countries, however, also show also policy-based partnership characteristics, especially in the USA and the UK (see also Chapters 6 and 7). In general, construction sector PPPs are merely project-based.

The typology of PPP contains prescriptions about the structure of the co-operation and the process of co-operation. The structure is the legal, financial, or

Table 2.1   Institutional aspects of public–private partnerships.

Financial aspects	— Project finance — Risk division — Revenue sharing
Legal aspects	— Contracts — Legal entity — Law and regulations
Administrative and organizational aspects	— Tasks and responsibilities — Project organization — Formal decision making requirements

organizational institution, whereas the process is the actual interaction. The structural aspects in PPP are summarized in Table 2.1.

Many studies have been undertaken on the financial, legal and contractual aspects of PPPs. Obviously these aspects are important conditions for success and cannot be ignored in a book on PPPs. However, as mentioned before, a key characteristic of PPP is the process of interaction in the projects. Most studies on PPPs do not pay attention to these interaction processes. This book explains the creation and functioning of PPPs by analysing the interaction processes between stakeholders.

Hence, the institutional nature and the interaction process form the basis of our definition of PPP:

■ Co-operation between public and private parties
  — At all stages of the project
  — In a project-specific organization
  — Involving all project risks
  — Under contractual arrangements
  — With contributions from all parties
  — With added value for all parties
  — With opportunities for generating cash flow.

2.2 General characteristics of PPPs

Understanding PPPs requires their general characteristics to be determined and defined. This section describes the characteristics of PPPs as an ideal model: the essentials that this specific organizational structure requires to generate the added value as intended.

Regarding the definitions, the essential elements of PPP are divided into four categories:

■ Actors (parties)
■ Network

- Project
- Relationship.

Nowadays, planning and establishing construction projects is a challenge for the public sector, the private sector and civil society. These three sectors each have their own (economic) identity and characteristics: the public sector is oriented towards public interest, social responsibility and environmental awareness; the private sector is thought to be creative and dynamic; civil society (the 'third sector') is strong in areas that require compassion and commitment of individuals (Rosenau, 1999). However, to an increasing extent, their activities and goals overlap (see, for example, Kickert et al., 1999; Dewulf & Spiering, 2001), with the shifting roles of government and the private sector blurring the lines dividing them. The 'new economic' thinking, for example, typifies this overlap as a 'mixed economy': greater interdependencies require more co-ordination across public and private organizational boundaries (Pongsiri, 2002).

If, when focusing on PPP, we regard the public and the private sectors as clusters of organizations, and organizations as assemblies of actors, it follows that public and private *actors* are to a certain extent interdependent when establishing their own organizational goals; for this reason the overlap can be typified as a *network* in which patterns of interaction exist between actors (Håkansson, 1989; Powell, 1991; Achrol, 1997; Peters, 1997). Defining a construction *project* to solve a certain problem does not require all patterns to be activated; only part of the network is activated for each specific problem. Trist (1983) calls the activated part of a network 'field-related organizational populations', noting that organizational populations become field-related when involved in a problem or a problem area that is a domain of shared concern. The linkages between actors, network and project are visualized in Fig. 2.1.

These field-related organizational populations can be organized in different ways to embody the shared concern. If a co-operative relation is formed to establish a specific project, the organizing can be characterized as an inter-organizational *relationship* (IOR) (Hellgren & Sternberg, 1995; Achrol, 1997). According to Gils (1978), an inter-organizational relationship can be either

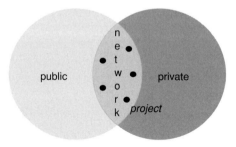

Figure 2.1   Project as an activated part of the public–private network (Bult-Spiering, 2003).

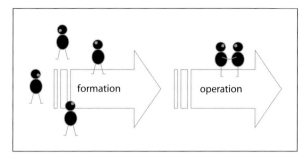

Figure 2.2 Public–private partnership as a temporary inter-organizational relationship in two phases.

temporary or permanent: if the relationship is started to establish a specific project, it will be temporary. A PPP in construction therefore is considered a temporary IOR, in the sense that it has a definable end-point.

The co-operation process in temporary IORs is divided into two phases: the formation phase and the operational phase. In the first phase, the temporary IOR is created and in the second it functions, as presented in Fig. 2.2.

Besides the distinction between the two phases, a difference between economic and sociological aspects of IORs is suggested in the literature which helps us to understand PPPs. In addition to more rational considerations (exchange of means, for example), more emotional considerations concerning the quality of the relationship itself are pertinent to the formation and operation of IORs and explain the creation of a specific IOR between certain actors. The economic and sociological aspects are *complementary*; both must be taken into account when describing and explaining IORs (Granovetter, 1985; Håkansson, 1989; Madhok, 1995; Foss & Koch, 1996).

Thus, to describe the characteristics of the ideal model PPP, we need insight into the economic and sociological essentials of the creation and functioning of temporary IORs, together with knowledge of the actors, network and other elements of the project.

2.2.1 Creating PPPs

This section starts by describing the characteristics of actors, networks and projects. These characteristics are merged to describe both economic and sociological characteristics of the relationship between public and private actors.

Actors

Actors' actions are determined by their own interests together with their perceptions of the interests of the other actors involved. Public and private actors have different goals, interests and organizational structures. To establish a successful relationship it is important to understand each other's goals and interests (Linder, 1999; Spiering & Dewulf, 2001).

Goals and interests

Both public and private actors try to make a profit; such profits are, however, of different types. Public actors aim to increase social revenues or welfare, whereas private actors aim to accrue business economic revenues. Welfare is influenced by the extent to which society's needs are met, but in the political marketplace there is no explicit relation between price or payment and the amount to which a certain need is satisfied (Lemstra et al., 1996). The orientation of the public sector is towards the following:

- Legislation, regulations and authorities
- Political opinion and political influence
- Democratic decision-making processes
- Minimization of risks
- Realization of a social goal (Reijniers, 1994).

Through the change towards governance and PPPs, different types of goods and services can be delivered with private sector involvement: delivering public goods is no longer an exclusively public task.

The starting point for determining the proper roles of government and the private sector is to classify the goods and services needed in modern society according to the degrees in which they are exclusive and in which they can be consumed. Four kinds of goods can be distinguished using this classification (Savas, 2000):

(1) Individual goods, such as food, clothing and housing (individual consumption and easy to exclude)
(2) Common-pool goods, such as fish in the sea (individual consumption, difficult to exclude)
(3) Toll goods, such as telephone, electric power, internet (joint consumption goods, easy to exclude)
(4) Collective goods, such as national defence, safety (joint consumption goods, difficult to exclude).

In the classic micro-economic approach, the private sector's goal of maximizing profits means minimizing the costs in the short term. The point in production where the marginal costs of expanding the production equal the marginal revenues is the business economic optimum with maximum profit. Equilibrium exists when profits are made, continuity is guaranteed, and when the ratio between incoming contributions and outgoing investments is favourable. In broader terms, the policy of firms is to try to achieve:

- Continuity of the firm
- Maximal return on investment.

The orientation of the private sector is towards the following:

■ Achieving returns on the invested funds
■ Taking business risks
■ Anticipating market and competitive developments
■ Realization of a corporate goal (Reijniers, 1994).

In the 'new' institutional economics and in economic-organizational theory, the existence of firms and inter-firm relations are, on a macro level, explained by property rights theory, transactions costs theory and principal-agent theory (Hazeu, 2000).

Public interest

In the political literature much attention is given to public interest. Substantial differences in definition or interpretation can be noticed. Cassinelli (1958), for example, considers the public interest as 'the highest ethical standard applicable to political affairs', while Braybrooke & Lindblom (1963) deny that public interest is the all-embracing aim of public policy, although they admit that in some cases this might be a helpful consideration. Others even argue that it is a meaningless concept in political theory. A review of the political literature indicates that, although the different perspectives offer interesting insights, an unambiguous descriptive definition of the public interest is not possible unless (quasi-) unanimity exists. The Dutch Scientific Council for Government Policy (Wetenschappelijke Raad voor het Regeringsbeleid, 2001) defines public interests as 'societal interests' if achieving them is entirely desirable for society. There is a public interest if the government feels responsible on the basis of the conviction that the interest is not otherwise looked after.

In economic theory, market failure is a reason to look at other institutional arrangements and the role of government.[1] Markets fail, for example, if not

---

[1] An in-depth discussion on aspects such as the price mechanism, quasi-collective goods, non-excludability and non-rivalry can be found in Cornes and Sandler (1986). An interesting question is, on what basis Smith pleaded for a very limited role for government and why Keynes pleaded for an extensive role. In his presidential speech 'The economist and the state' to the American Economic Association in 1964 Georg Stigler asked why the insights of economists had changed so drastically over the years, indicating that economists had not made a serious study of costs and benefits of different institutional arrangements, the different perspectives were not scientifically underpinned and that economists only followed the tendencies in society. He called for systematic scientific research to ascertain the optimal interaction between market and state which led to the gathering of much information. Van Damme (2001) maintains that these results lead to the conclusions that private parties perform better than public parties and competition improves performance. In his opinion, this convincing evidence has clear implications for policy: the use of markets and privatization is very likely to be the best way to safeguard public interests. The motto is 'yes, on condition that' and 'no, unless', the condition being whether the government can properly fulfil its role as 'market-director'.

all positive and/or negative (external) effects are expressed in the price. In other words, we refer to market failure if an action of one party or the transaction between two (or more) parties has positive or negative consequences for other stakeholders who are not directly involved in the action or transaction. The consequence is that the private interests do not coincide with the interests of society, because not all external effects are included in the deliberation. If external effects are complicated such that many parties are involved, additional (trans)actions are hard to realize because of the problem of 'free-riders'. In these cases there is a public interest that can be satisfied by public intervention. In contrast, interference by the government could have negative implications. The 'public choice theory'[2] offers several arguments for 'government failure'. Government intervention could lead to a division between costs and benefits which might result in higher costs, the internal goals of an organization might replace public interests, and actions of the government could also cause anticipated and non-anticipated external effects.

Economic or public choice theories do not reveal a single answer concerning public interest and the role of government in safeguarding it. The discussions could be characterized as normative (political) deliberations.

In most publications, the discussion about public interest focuses on the structural allocation or rearrangement of tasks and responsibilities between public and private parties in so-called network-sectors, such as energy distribution and public transport (Raad voor het Openbaar Bestuur, 1998; Ministry of Economic Affairs, 1999; Wetenschappelijke Raad voor het Regeringsbeleid, 2001). The debate concentrates on what institutional system or arrangement is best able to safeguard the public interest. This book concentrates on new forms of governance at project level. Little research is done, however, on the way public interest can be safeguarded at the project level.

Roles

The structure of the different types of public actors (central, regional and local government) is a layered one: this layered structure concerns the division of tasks and responsibilities, and mutual relations. In construction projects, all types of public actors are involved, though in a different way, both during the policy-making process and during implementation of those policy decisions. Also,

---

[2] Public choice theory studies decision processes with the use of formal analyses on the basis of the so-called homo-economicus postulate. This postulate has no empirical content. It means nothing more than that individuals try to realize their own interest in the best possible way. This is expressed in 'systematical behaviour'. The motives by which individuals are driven is left open. Because the analytical instruments from economy are applied to political and administrative phenomena, the 'public choice' theory has an interdisciplinary character (Klaver & Siccama, 1974). The approach has become well known to authors such as Black, Arrow, Downs, Buchanan, Tullock and Olson.

within each type of public actor there is a layered structure between politicians and civil servants (policy sectors). The complicated games of interest often make the layered structure problematic.

The private actors which can be involved in planning a construction project are subdivided according to the type of tasks they fulfil: designers, developers, constructors, financiers and investors.

Networks

Where construction projects are involved, the activities of public and private actors overlap, interaction patterns evolve giving a network:

> 'Networks are changing patterns of relations between mutually dependent actors, that organize themselves around problems or means.' (Teisman, 1998)

This network structure is often found not only in construction projects, but also in other fields of governance. The interlinking between public and private sectors is an important cause of network origination. Networks are, in fact, structures or systems in which certain patterns of interaction are recognized. In this book, we focus on networks between actors of different organizations, and not on networks within organizations.

The network approach is often seen as a third mechanism for the co-ordination of economic activity in addition to the two forms distinguished by the transaction costs theory of Williamson (1983): market and hierarchies. The most important criticism of Williamson is that these two structures do not take the meaning of personal relations, reputation and trust into account (see, for example, Douma & Schreuder, 1998). Powell (1991) compares three forms of economic organization: market, hierarchies and networks, on the basis of seven characteristics (Table 2.2).

In public sector literature, the network perspective is often described as an alternative for conventional governance models such as the rational central rule perspective and the multi-actor perspective (see, for example, Kickert et al., 1999).

Networks are of a multiform nature: many different actors are present. These actors have complementary activities and qualities as well as mutual interdependencies; this strengthens the commitment between actors and their striving for mutual revenues (Håkansson, 1989; Powell, 1991; Alter & Hage, 1993; Achrol, 1997; Peters, 1997; Teisman, 1998; Bruijn & Heuvelhof, 1999; Laws et al., 2001). Actors in networks have non-hierarchical, long lasting relations; therefore networks are, as structures or systems, clusters of interdependent organizations and of inter-organizational relationships (Breuer, 1978; Gils, 1978). These relationships are activated for solving specific problems. Not every problem asks for the activation of the same relations in the network. For each problem, an appropriate part of the network is activated. Networks therefore are highly dynamic: actors enter and depart, power positions change and different networks can come

Table 2.2   Stylized comparison of forms of economic organization (based on Powell, 1991).

Key features	Market	Hierarchy	Network
Normative basis	Contract – property rights	Employment relationship	Complementary strengths
Means of communication	Prices	Routines	Relational
Methods of conflict resolution	Haggling – resort to courts for enforcement	Administrative fiat – supervision	Norm of reciprocity – reputation
Degree of flexibility	High	Low	Medium
Amount of commitment among the parties	Low	Medium to high	Medium to high
Tone or climate	Precision and/or suspicion	Formal, bureaucratic	Open-ended, mutual benefits
Actor preferences or choices	Independent	Dependent	Interdependent
Mixing of forms	Repeat transactions Contracts as hierarchical documents	Informal organization Market-like features: profit centres, transfer pricing	Status hierarchies Multiple partners Formal rules

together. Because of this dynamic nature, the reputation of the actors is important: relations with actors having a negative reputation will not be activated to the same degree as relations with actors having a more positive reputation (Håkansson, 1989; Godfroij, 1992; Hellgren & Sternberg, 1995; Bruijn et al., 1998).

Projects

Based on project management literature Packendorff (1995) states that a project is usually defined as:

■ A unique, once-in-a-lifetime task
■ With a predetermined date of delivery
■ Being subject to one or several performance goals (such as resource usage and quality)
■ Consisting of a number of complex and/or interdependent activities
■ A project is therefore a given, plannable, and unique task, limited in time, complex in its implementation and subject to evaluation.

Turner's (1993) definition is similar:

> 'A project is an endeavour in which human, material and financial resources are organised in a novel way, to undertake a unique scope of work, of given specification, within the constraints of cost and time, so to achieve beneficial change defined by qualitative and quantitative objectives.'

Hellgren and Sternberg (1995) link projects to interaction, and consequently relate the characteristics of a project to the characteristics of actors and network:

> 'A project is seen as processes of organizing among actors with different rationalities in terms of goals, time orientation and problem solving, and different and changing power positions in the network.'

The establishment of construction projects demands many resources, which are scarce. These resources include financial means as well as deployment of knowledge and capacity. However, construction projects also create positive effects for their surroundings. Improvements in the economic and investment climate often generate regional spin-off and valued projects have an important positive effect on the reputation of the actors involved.

Creating a relationship

Economic aspects

A transactional motive for interaction is the exchange of resources (Williamson, 1983). When actors need knowledge, capacity and financial means, IORs are likely to arise (Alter & Hage, 1993). The exchange of resources gives the actors an economic advantage if the resources are complementary (Peters, 1997). In addition, external pressure from the direct environment of the actor's organization can be a motive for co-operation. Both motives, exchange and external pressure, are fortified by competition with other organizations in the supply of certain products (Andriessen, 1989).

Chances for co-operation increase when the co-operation creates added value (Lemstra, 1996; Bruijn et al., 1998). Before participation, private actors judge projects on profitability and the way in which the project goal meets the organization's internal goals. The same is true for public actors, with the addition that profitability shows itself in social revenues. After this judgement, selection of the actors with whom to co-operate takes place. In most cases, the public actor selects the private actor, although private actors also have their own requirements regarding their public partners. Most construction projects also have to meet specific procurement rules. In some cases, private actors have a landowner's position in the project: this means the public actors are forced at least to negotiate with certain private actors.

Through PPP, public and private actors expect to create added value: the co-operation leads to results that could not be achieved by the parties acting alone. Added value appears in four forms:

(1) *Added value in content* This is accomplished at the project level by an integrated approach to problem solving and the realization of different, coherent functions. PPP improves the quality and innovation of the solution, e.g. the actual project

(2) *Added value in process* This is effected by the early combination of complementary knowledge and experience, and by adjusting goals and interests. Private actors are expected to react more effectively and efficiently than can public actors, they have greater financial strength and knowledge of the relevant markets. Likewise, private actors can use the public actor's knowledge and competences concerning political procedures and decision-making processes

(3) *Financial added value* This is achieved by the division of risks and making adjustments to give an improved price/quality ratio. For example, public spaces or infrastructure become affordable by returns on real estate developments, while returns on real estate developments are increased by high quality public spaces and increased accessibility. Through co-operation, risks can be spread and therefore reduced per actor

(4) *External added value* This is effected by the co-ordination of different projects and initiatives. Developments in a certain area are often threatened by developments in adjacent areas. Through co-operation in PPPs, private actors can influence public activities in these developments and different initiatives can be harmonized.

Early involvement of the private sector generates the optimum possibilities for exchange and added value. Savas (2000), for example, describes how the private sector can help the government to deliver infrastructure services:

■ Identify and develop new, innovatively designed, user-financed, profit-making facilities or existing facilities needing rehabilitation, renovation or expansion
■ Private sponsorship and commercial loans
■ Access to private capital markets to supplement or replace public financial resources
■ Satisfy public needs more quickly and at lower cost
■ Operate facilities more efficiently
■ Provide new sources of tax revenue
■ Accept risk.

Sociological aspects

Motives for starting IORs are also concerned with non-rational aspects of co-operation. Alter & Hage (1993) state that, next to transaction-based motives, willingness to co-operate is an important aspect of creating temporary IORs. According to Andriessen (1989), mutual consciousness of the existing interdependencies is also necessary for starting IORs; co-operation can be rationally attractive. However, when willingness and consciousness are missing it will be difficult to establish a PPP (Peters, 1997).

A third sociological aspect in creating IORs is related to organizations' ideologies. Those organizations which have policies with a clear and strong focus on co-operation will be more likely to choose to create an IOR than inward-oriented organizations.

Mutual perceptions of each other's starting points, goals, interests, possible share, competence and reliability are a fourth sociological category when creating IORs. Positive expectations of the possible partner regarding these aspects make the creation of IORs more likely. Not only are the perceptions and expectations of the possible collaborator important, but also perceptions and expectations of their own organization: effects on their own image or identity, and the perception of their ability to co-operate in view of their organization's structure and culture, are mentioned in this respect (Andriessen, 1989; Bruijn et al., 1998).

In addition, the actors' fit, i.e. the degree to which the actors' characteristics correspond and influence their willingness to co-operate, is important in creating IORs. The relevant characteristics are organizational structure, culture, goals, operating philosophy (operating methods and vision), level of expertise and ideological consensus (accord in values and standards) (Andriessen, 1989; Schultz, 1994). Finally, domain consensus, which exists when actors recognize each other's right to be active in a certain domain or field of work, is a sociological aspect that can explain why temporary IORs are created (Andriessen, 1989).

2.2.2 The functioning of PPPs

The creation of an IOR is followed by the actual functioning of the IOR when a distinction is made between economic and sociological aspects that describe and explain the operation. Economic aspects deal with the goal (rationale), type and structure of the partnership, while sociological aspects concern the behaviour of the actors in the relationship. This section first describes relevant aspects of actors and projects. The nature of the public–private network is primarily relevant for the creation of PPPs, but not directly for their functioning. The network perspective enables us to identify a project's stakeholders, with their interests and resources. Economic organizational theories help us understand the co-ordination mechanisms underpinning the actual co-operation and its formalization in contracts. Networks influence the functioning of PPPs in a context-related way, but are not interactions characteristic of the partnership itself.

Actors

In striving to achieve their goals, public actors try to minimize risks, whereas private actors will have to take risks to make profits, especially in a highly competitive market. Public organizations are open systems in which all kinds of stakeholders influence the decision process. This process is a layered one involving different types of actor. The public decision-making process is based on political preferences expressed, either directly or indirectly, by the electors through the voting mechanism. Competition, however, determines the price of goods produced on the private market. In most cases, consumers can choose between many products and producers. A firm has to be aware of consumer preferences and behaviour as expressed through the pricing mechanism. Business decisions are based on reconciling prices on the input and output sides of the production process, such decisions being made within a relatively closed system (Lemstra et al., 1996).

The stage in the project at which an actor's interests become visible differs for each type of private actor: designers and developers are active in the development and planning of a project, whereas constructors are active in the realization phase. Financiers and investors are merely interested in the final product and its revenues; hence they often have a position relatively remote from the actual project.

Many other actors, such as private citizens, interest groups, estate agents, architects and consultants, are involved in construction projects. These actors are of contextual importance and their roles therefore will not be considered here.

Projects

Projects are temporary and aim to achieve clearly defined goals to solve a specific problem; when the project goals have been achieved the project is finished. Nevertheless, the realization of construction projects takes a long time because they have high political and social priority, which makes political and social support essential; decision-making processes are often extensive.

Furthermore, construction projects are confined to a specific location, the resources needed are scarce, and in most cases already in use; consequently, many actors are involved. Ultimately, construction projects have a high risk profile. Risks are caused by developments within the project and also by exterior factors. Project feasibility is influenced by local, regional and national market developments and by the actors' activities in other projects. Jacobs (1998) identified the following risks in partnership projects:

- Procedural risks due to changing public policy or not obtaining consents or licences
- Design risks due to problems resulting from the process or project design
- Construction risks due to time or cost overruns or defective construction
- Maintenance risks due to problems that arise during the maintenance of an asset over time
- Operating risks due to ineffective operation of an asset and therefore failure in meeting its target
- Financing risks due to changing economic conditions
- Revenue risks due to the disappointing performance of the asset and consequently its insufficient generation of income.

Li Bing et al. (2005) distinguish specific concession PPP risks at three levels:

- Macro level risks including political and government policy risks, macro-economic risks, legal risks, social risks and natural risks
- Meso level risks including risks concerning project selection, project finance, design, construction, operation and residual risks
- Micro level risks including relationships and third party related risks.

Relationship functioning

Economic aspects

Madhok (1995) refers to the economic aspects of IORs as the 'contract-centred approach': the project organization's structure and the chosen type of contract are the focus. As stated before, a temporary IOR is an activated part of a network for the solution of a certain problem or the realization of a specific project and is hence a knowingly started mutual commitment between two or more organizations to achieve a certain goal (Luscuere, 1978). Achrol (1997) uses the term 'opportunity networks' to emphasize the finite character of the partnerships and the aim for a set goal. These goals must be set down and elaborated in a concrete plan (Rosenau, 1999).

Although projects are limited in time, their establishment is often a lengthy process, so they are subdivided into phases. Furthermore, each project requires a specific organizational structure (Reijniers, 1994). The co-operation between actors can involve all project phases, but can also be limited to one or a few phases. For each phase the necessary contributions are determined and which project organization matches these contributions: private actors could contribute their promptitude, market knowledge and experience, networks, risk-bearing capability, and insight into the project's feasibility from a market perspective; possible contributions of public actors in construction projects consist of knowledge, their influence on public law, regional market knowledge, reliability, networks, funding capability and insight into the project's feasibility from a political perspective.

Sociological aspects

The way the parties interact influences the functioning of inter-organizational relationships. Madhok (1995) refers to this sociological perspective as the 'relationship-centred approach'. The following sociological concepts can be identified (Gils, 1978; Andriessen, 1989; Schultz, 1994; Achrol, 1997; Peters, 1997):

- Commitment
- Flexibility
- Perseverance and leadership
- Trust
- Acceptance and respect.

Table 2.3 General characteristics of prototype PPP (based on Bult-Spiering, 2003).

Characteristics		Actors		Network	Project
		Public	**Private**		
Creating PPP	Economic	**Achieving the organization's goals (project goal and spin-off)**		Many different actors	**Exchange of means because of scarcity**
		— Revenue (social)	— Revenue (economic)	Complementary means, qualities and activities	— **Financial means**
		— Efficiency	— Efficiency	Mutual dependencies	— **Capacity**
		— Control	— Shareholders	Reputation is important	— **Expertise**
		— Public interest		External pressure	**Added value**
			Roles	Competition	— **On content**
		— Central, regional, local government	— Designer, developer, builder, lender, private contribution, investor	Strong dynamics	— **On process**
		— Civil servants, politicians			— **Financial**
		— Public contribution	— Private contribution		— **External**
	Sociological	**Willingness to co-operate and awareness of mutual dependencies (willingness can be part of organization's ideology)**		**Long lasting non-hierarchical patterns of relations**	Domain consensus
		Positive expectations and perception of			
		— Each other's starting points, objectives, interests, possible share and reliability			
		— One's own image/identity and ability to co-operate			
		Actors' fit			
		— **Structure**			
		— **Culture**			
		— **Goals**			
		— **Operating philosophy**			
		— **Level of expertise**			
		— **Ideological consensus**			
Functioning PPP	Economic	Minimizing risks	Taking risks		Solving a certain specific problem, with high priority and social relevance
			Traditional organization		Bounded to location, unique, limited in time, indivisible, large risks
		— Decisions by many actors	— Decisions by limited number of actors		Contracts and project organization for each kind of project and each project phase
		— Voting mechanism	— Pricing mechanism		Careful internal and external communication and social support
		— Open system	— Closed system		
		Achieving collective goals and mutual revenues			
	Sociological	**Internal relationships**			
		Knowingly started public–private relationships			
		Dynamics caused by limited mutual dependency			
		Behaviour			
		— **Commitment**			
		— **Flexibility**			
		— **Perseverance**			
		— **Trust**			
		— **Acceptance and respect**			

The behaviour of the actors towards each other can be explained by these concepts. The success of a partnership will depend on the level of commitment for an extended period and also a flexible attitude to the operating procedures and framework of the actor's own organization. Trust and joint image are important elements in the creation and functioning of PPPs because all projects are confronted with uncertainties that cannot be banished by means of contracts (Koppenjan, 2005). Many authors have stressed the importance of interaction in inter-organizational co-operation (see, for example, Håkansson & Johansson, 1993). Furthermore, because co-operation must be maintained for a long time, perhaps over several city council election periods, a certain amount of perseverance is needed, a so-called stayer's attitude. In addition, leadership must be available to public and private actors, and both must be able to negotiate effectively with each other (Peters, 1997).

Partners need to trust each other's professionalism and reliability. Building trust is, however, a cyclic process which is why the process of interaction needs 'nurturing' (Vangen & Huxham, 2003). In addition, acceptance and respect concerning positions, roots and problems are crucial; therefore there has to be a social–emotional bond between the co-operating actors.

During co-operation, the behaviour of the actors is aimed at both establishing the collective project goal and their own organization's goals: these goals are usually compatible, but can be contradictory (see, for example, Stephenson, 1991) in which case the actors are no longer interdependent in achieving their goals. The mutual dependency of actors in temporary IORs is therefore limited and it is this which gives the IOR a dynamic character. Thus, not only the network in which the IORs are created, but also the IORs themselves are dynamic (Gils, 1978; Godfroij, 1992; Austin & McCaffrey, 2002). The general characteristics of interaction in the ideal model of a PPP are summarized in Table 2.3.

In practice, not all the ideal model characteristics are present in PPPs, as evaluation shows. In fact, certain elements of the framework presented in Table 2.3 in bold type are critical for the effective performance of PPPs. These are further elaborated in Section 2.4 and form the framework of analysis in this book.

2.3 Construction sector PPPs

Although PPPs are nowadays applied in diverse areas such as defence (including base operations, technology, infrastructure and logistic services) and the environment in developing countries (United Nations Development Program Public–Private Partnerships for Urban Environment) where PPPs are used to improve the delivery of urban environmental services, in this book we focus on PPPs in the construction sector. PPPs are used in many construction sector fields. The two fields in which PPPs appear to be most dominant throughout the world are urban development and infrastructure development. Public infrastructure is vital to a nation's production and distribution of private economic output as well as to its citizens' overall quality of life. 'Hard' or 'economic' infrastructure commonly comprises roads, transport systems, communications, water and sewerage, electricity, gas and ports; 'soft' or 'social' infrastructure comprises schools, universities, research facilities, military housing, waste water treatment plants, prisons, hospitals, libraries, public buildings and parks.

2.3.1 Urban development

Urban development examples of PPPs are usually mixed-use developments within certain areas; these typically include housing, retail space, office space, hotels, parking, garages and public space. For these mixed-use developments, joint venture PPPs dominate.

> An example of PPP in urban development can be found in Bercy in eastern Paris (Nelson, 2001). The Mayor of Paris owned the land before development and, as he effectively was the planning authority, redeveloped the site of the former warehouses at Bercy by creating a 'zone d'aménagement concerté' (ZAC). This implied that the Bercy area was scheduled for comprehensive redevelopment under a special planning regime, where revenues from sale of land are used for infrastructure and public facilities. The responsibility for the implementation of the redevelopment was delegated to a 'Societé d'Economie Mixte' (SEM) for eastern Paris (SEMAEST). SEMAEST is jointly owned by the Mayor of Paris, the nationalized state railway, 'Societé Nationale des Chemins de Fer Français' (SNCF) and 'Societé Anonyme de Gestion Immobilière' (SAGI). SAGI is jointly owned by the Mayor and a private property company.

2.3.2 Transport infrastructure

Transport infrastructure examples of PPP are merely single use developments of roads, (light) railways and transit systems. The concession type of PPP is commonly used.

> An example of a transport infrastructure PPP is HSL-Zuid in The Netherlands, concerning the design, realization, maintenance, financing and operation of 100 km of railway (superstructure) and systems for high-speed trains. The project comprises railway construction, energy systems, control and security systems, acoustic fencing and telecommunications. The contract value is €1.3 thousand million and the private consortium is Infraspeed.

2.3.3 Social infrastructure

Social infrastructure PPPs involve the single use development (although different uses of real estate may be combined, the development of different categories of spatial function such as infrastructure and open spaces is not integrated) of facilities, including construction of new facilities, rehabilitation of existing facilities, expansion of existing facilities, or demolition of existing facilities. Just as in transport infrastructure, concession PPPs are commonly used.

An example of a social infrastructure PPP is Her Majesty's Prison Altcourse. This was the first prison to be designed, constructed, managed and financed privately in the UK. Group 4 was awarded the contract in 1995 and building commenced in partnership with Tarmac Construction. The prison opened its doors to prisoners on 1 December 1997, some six months earlier than the original anticipated opening date. The prison employs a team of over 400 high quality local people, all working to the same aim of producing excellence in custodial practice. HMP Altcourse is a Category A core local prison receiving young offenders and adult male prisoners from the courts in Merseyside, Cheshire and North Wales. The prison currently holds over 900 prisoners in safe custody.

2.4 Problems in PPP practice

Public–private partnerships are considered an important tool for realizing construction projects. The nature of the problems, the need for an integrated approach, and the fact that activities of private and public actors overlap require non-traditional operating procedures. However, the benefits of PPP are still disputed despite the many advantages.

Advocates of PPP emphasize the added value of private involvement in construction projects. By working more closely with business, governments can use market forces. Efficiency, knowledge and customer orientation are market strengths that can improve public sector performance and products. In addition, PPPs stimulate the private sector to undertake life-cycle approaches and supply chain management, which are said to improve the price/quality ratio. Early private sector involvement in construction projects makes this sector more aware of the project's benefits for society.

In practice, however, PPPs often do not involve the intensive co-operation required in the ideal model. In some cases, the partnerships even border on privatization (Rosenau, 1999; Dewulf et al., 2004), causing failure in the performance of the PPP, and therefore often a negative connotation with the idiom. Public–private partnership is not always the right solution: it is applicable in certain situations to meet a particular set of problems and circumstances. Theory is necessary to match PPP as an instrument with situations (Peters, 1997), which is why we developed the framework described in Section 2.2. To use PPPs in an adequate way in specific situations, and thereby utilize their advantages instead of their disadvantages, we need to face the threats they pose.

In general, these threats account for both the US and the European contexts and based on literature and empirical research, are subdivided into two categories of failing performance: *product* and *process* performance. Although Alexander (2000) states that systematic analysis of interaction performance in urban development is problematic and generalization is difficult because of the complexity and multitude of relevant factors, several conclusions on performance can be drawn within the context of PPPs. Differences between PPP ideal model characteristics and PPP practice are examined in the next section. As the roots for both types of performance lie in the assumed added value of PPPs this will also be a returning theme in the more project-specific Chapters 4 to 7.

2.4.1 Product performance

Product performance concerns the actual goals and the delivered product(s) of the PPP; it can be split into two categories:

- Financial performance (cost efficiency, transaction costs, risks)
- Content performance (value for money, innovation).

Financial performance: cost efficiency, transaction costs and risks

The evidence on financial performance advantages of PPPs is still mixed, even though externalities are seldom considered in analysis (Rosenau, 1999). Comparing PPP contracts to public sector enforcement is difficult because of the lack of quality data on, for example, maintenance costs and established performance.

According to Rosenau (1999), PPPs may be appropriate if cost is the main concern, externalities are expected to be limited, and a relatively short time-frame is in place. Public–private development involves all partners in time-consuming interactions; PPP procurement therefore has relatively high costs and takes a long time (Stoker, 1997; Stainback, 2000; Knowledge Centre Public–Private Partnerships, 2004). In many cases, the selection of the private consortium is a lengthy process involving high private participation costs, with much negotiation and management time spent in the contract transactions (Ezulike et al., 1997; Li Bing & Akintoye, 2003). In addition, the long duration of concession contracts demands flexibility, and dealing with changes is a significant and often problematic issue (NAO, 2001).

The lower costs of public loans are often used as an argument against the presumed financial performance advantages of PPPs because public sector completion of projects can often be cheaper than bringing in the private sector (Bult-Spiering et al., 2005). Clarke (1998) and Stoker (1997) mention the transaction cost dilemma in this context. Due to economic and political dynamics and increasing social complexity, co-operation is a necessary precondition for lowering the transaction costs; however, because trust and similar values are required, co-operation results in high transaction costs.

Another related aspect of financial added value of PPPs is the allocation of risks. In joint venture PPPs, risks are divided between public and private participants, whereas in concession PPPs risks are transferred to the private sector. In both cases, defining and quantifying the risks involved in these projects is difficult and complex (Li Bing & Akintoye, 2003; Li Bing et al., 2005). Even the actual transfer of risks is suspect (Gaffney et al., 1999).

Content performance: value for money and innovation

Public–private partnerships are supposed to add value in terms of value for money and innovation; this applies particularly for concession PPPs. Although in most countries these two aspects are the main motives for starting PPPs, actual enhancement of value for money and innovation through PPP is not always provable. Specifically, the contribution of PPP to design and planning innovation is questionable.

An important instrument for quantifying the value for money of a partnership for a particular project is the public sector comparator (PSC). The PSC is used to calculate the 'in-house' costs of delivering a project to determine if a PPP demonstrates good value for money. This instrument is, however, often criticized for its rigid use and the problem of comparing more qualitative, intangible aspects (NAO, 2001; Dewulf et al., 2004).

Despite the focus on value for money and innovation, especially in concession PPPs, 'low costs' are still used as a selection criterion; this eliminates the possibility of making selections based on 'best value' where performance and costs are balanced (Gansler, 2003; Dewulf et al., 2004). Awarding contracts through competitive bidding decreases the possibilities for co-operation in initial planning discussions. Both public and private partners seem unable to find sufficient incentives in PPPs to take risks through innovation (Hall, 1998).

2.4.2 Process performance

Process performance concerns the actual co-operation in the partnership and the way in which the product of the PPP is realized. Process performance is characterized by high dynamics: within a certain PPP process, the extent to which certain sociological factors, such as trust and willingness to co-operate, are present differs in time. Also, the previous and future interactions between parties (the sociological context) influence the interchanges in a certain PPP process.

Here we distinguish three issues of process performance:

- Actors' fit and willingness to co-operate
- Public interest
- Behaviour.

Actors' fit and willingness to co-operate

Challenges to PPPs caused by suboptimal utilization of partnerships are to find ways of dealing with internal disharmony. This disharmony can exist both within the partnership and within the private or public sector itself (Austin & McCaffrey, 2002). Disharmony within the partnership is caused by conflicts of interest (Rosenau, 1999). Disharmony within the sectors themselves is most dominant in the public sector because the different levels of government and different policy sectors involved are not co-operating efficiently (Bult-Spiering, 2003; Knowledge Centre Public–Private Partnerships, 2004). Many literature reports suggest that the co-operation between the two different sectors often leads to negative outcomes, such as increased complexity, loss of decision-making autonomy and information asymmetry (Williamson, 1983; Pongsiri, 2002). Regeneration processes in particular, are long and complex (Klijn & Teisman, 2003). Communication between public and private sectors is often not clear: both sectors speak different languages, and have different experiences and expectations (Stainback, 2000; Bult-Spiering, 2003). Tensions exist because of the differences between public and private actors. Table 2.4 gives an overview of public and private actors' characteristics and possible tensions originating from the dissimilarities.

Clarke (1998) also mentions the governance dilemma of PPPs:

> 'Due to greater complexity, cooperation is needed among multiple actors, but not too much: generating "enough cooperation" among those possessing the strategic resources and knowledge is necessary to pursue a specific agenda.'

Thus, a tension exists between the need for co-operation because of increased complexity on the one hand, and the increased complexity resulting from the emergent co-operation and actors involved on the other.

An additional dilemma concerns stability. Public–private partnerships need continuity in policy and the ability to adapt to changing circumstances, i.e. to be flexible (Fosler & Berger, 1982). Because public terms of office are mostly much shorter than the time needed for project delivery, this continuity is challenged. Furthermore, PPPs often lack the required endurance of relationship: the partnership is often transformed into 'a set of loosely-linked projects' (Klijn & Teisman, 2003), and the absence of shared responsibility (Rosenau, 1999).

Public interest

No clear answer can be given to the question, what is the best way to define and safeguard the public interest? The choices made will, to an important extent,

Table 2.4 Characteristics of public and private actors and potential tensions in PPP (Klijn & Teisman, 2003).

	Public actors	Private actors	Tension
Core business	Objectives: (sectoral) public objectives Continuity: political conditions	Objectives: realizing profits Continuity: financial conditions	Different problem definitions: political risks in expectations versus market risks in annual figures
Values	*Guardian syndrome* Loyalty Devoted to a self-defined public cause Controllability of process and approach (political/social) Emphasis on risk avoidance and preventing expectations	*Commercial syndrome* Competitive Devoted to consumer preferences Controlled by shareholders on the basis of results Emphasis on market opportunities, risks and innovations	Government's reluctance in process versus private actor's reluctance with knowledge Government's reluctance in result versus private parties' reluctance with their own effort
Strategies	*Task organization* Search for ways to guarantee substantive influence (primacy of the public sector) Minimizing expectations and insecurity of implementation costs	*Market organization* Search for certainties to produce and/or obtain a contract. Minimizing political risks and organizational costs as a consequence of public 'viscosity'	Confrontation leads to a mutual 'locking-up' of agreements and thus to tried and tested types of co-operation (contracts)
Consequences for PPP	Emphasis on a limitation of risks and on agreements that lead to agreed procedures and public sector dominance	Emphasis on certainty of market share and profit, which leads to an expectant attitude and limited investments until the moment when the contract is acquired	The creation of added value through cross-border interaction is not realized

depend on the user's perspective and the characteristics of the sector. Accordingly, at the start of a construction project, it is not possible to define a common 'public interest' that contains measurable performance indicators for the final result. Decisions about how to safeguard the public interest have to be made in the light of their context and have to be able to cope with all possible changes. It is not sufficient to formulate process criteria such as transparency, openness and legitimacy. A process is not self-justifying, but is aimed at some substantial result. So, in practice, the (normative) definition of the public interest will contain both process and product indicators. Although criteria for the process may be determined at the first stage, in such projects this is not possible for (detailed) criteria concerning the end result. The point of departure, however, should be that both product and process approaches by themselves are not sufficient, but that they should be combined.

Adversaries of PPP fear that they endanger public interest. The private sector's profit-seeking goals are supposed to conflict with public values (Bult-Spiering et al., 2005). That partnerships may be the overall solution for problems that cannot be 'efficiently' solved by the public sector alone is a myth. The private sector is presented as innovative, skilled and efficient; community values, in particular, are threatened by this supposition (Stephenson, 1991; Clarke, 1998). New public management (NPM), in blurring the public–private distinction, could also be dissolving accountability, transparency and democratic choice. In particular, the ability of project-based partnerships to solve certain public sector problems is questionable. In PPPs, the public sector must trade some of its autonomy and use of its authority for the co-operation of the private sector: this is a potential danger for public interest and public accountability (Peters, 1997). Unclear goals and responsibilities, especially, reduce policy accountability (Rosenau, 1999; Stainback, 2000). In addition, problems with regard to equity, access, participation and democracy, challenge the public interest performance of PPPs (Rosenau, 1999). Finally, PPPs do not seem to reduce regulation; in some cases they even increase it (Rosenau, 1999).

Behaviour

Behavioural aspects such as trust, respect and flexibility play an important role in partnerships. As stated previously, trust, for example, also influences the transaction costs of the partnership. Interactions between public and private parties highlight the importance of these behavioural aspects. Practice shows, however, that the strength of public–private relations, i.e. the basis of the behavioural aspects, traditionally varies between countries (see, for example, Spiering & Dewulf, 2001; Dewulf et al., 2004). In addition, in certain national construction sectors, little attention is paid to the relational aspects of business transactions and the focus is on contractual arrangements (Egan, 1998; NAO, 2001; Dorée, 2004).

2.5 Summary

Public–private partnerships appear in many forms and are defined in various ways. In this chapter, the general characteristics of the ideal PPP are outlined and a general definition derived. These characteristics concern the following categories:

- Actors
- Networks
- Projects
- Relationships: economic and sociological aspects of creating and functioning of PPPs.

Although PPPs are considered important in many sectors of the construction industry, problems arise concerning the product and process performance of such partnerships. These problems are outlined in this chapter and will be further specified in Chapters 4 to 7. Each of these chapters focuses on a different construction sector in which PPPs occur, as outlined above. Chapter 3 describes the formal context of public procurement that the two main types of PPPs, concessions and joint ventures, face in the European and American context.

References

Achrol, R.S. (1997) Changes in the theory of interorganizational relations in marketing: toward a network paradigm. *Journal of the Academy of Marketing Science*, **25** (1), 56–71.

Alexander, E.R. (2000) Inter-organizational coordination and strategic planning. In: W. Salet & A. Faludi (Eds) *The Revival of Strategic Planning* (pp. 159–174). Amsterdam: Royal Netherlands Academy of Arts and Science.

Alter, C. & Hage, J. (1993) *Organizations Working Together*. Newbury Park: Sage.

Andersen, O.J. (2004) Public–private partnerships: organizational hybrids as channels for local mobilisation and participation? *Scandinavian Political Studies*, **27** (1), 1–21.

Andriessen, J.H.T.H. (1989) Organisaties en hun relaties. *IVA Methodencahiers*, **6**, 26.

Austin, J. & McCaffrey, A. (2002) Business leadership coalitions and public–private partnerships in American cities: a business perspective on regime theory. *Journal of Urban Affairs*, **24** (1), 35–54.

Baybrooke, D. & Lindblom, C. (1963) *A Strategy of Decision*. New York: Free Press.

Beauregard, R.A. (1997) Public–private partnerships as historical chameleons: the case of the United States. In: J. Pierre (Ed) *Partnerships in Urban Governance: European and American Experiences* (pp. 52–70). London: Macmillan.

Breuer, F. (1978) De interorganisationele analyse. *M&O*, **1**, 32–44.

Bruijn, J.A. de & Heuvelhof, E.F. ten (1999) *Management in Netwerken*. Utrecht: Lemma.

Bruijn, J.A. de, Heuvelhof, E.F. ten & Veld, R. in 't (1998) *Procesmanagement*. Schoonhoven: Academic Service.

Bult-Spiering, M. (2003) *Publiek–Private Samenwerking: de Interactie Centraal.* Utrecht: Lemma.

Bult-Spiering, M., Blanken, A. & Dewulf, G. (2005) *Handboek Publiek–Private Samenwerking.* Utrecht: Lemma.

Cassinelli, C.W. (1958) Some reflections on the concept of the public interest. *Ethics,* **69** (1), 48–61.

Clarke, S.E. (1998) Economic development roles in American cities: a contextual analysis of shifting partnership arrangements. In: N. Walzer & B.D. Jacobs (Eds) *Public–Private Partnerships for Local Economic Development,* 19–46. Westport: Praeger.

Cochran, C.E. (1974) Political science and 'the public interest'. *Journal of Politics,* **36** (2), 327–355.

Cornes, R. & Sandler, T. (1986) *The Theory of Externalities, Public Goods and Club Goods.* Cambridge: Cambridge University Press.

Davis, P. (1986) *Public–Private Partnerships: Improving Urban Life.* New York: Academy of Political Science.

Dewulf, G. & Spiering, W.D. (2001) Public–private partnership: the difference between innercity and infrastructure projects. *7th International Conference on Public–Private Partnerships.* Enschede: Enterprise Governance.

Dewulf, G., Bult-Spiering, M. & Blanken, A. (2004) *Opportunities for PFI in The Netherlands.* Enschede: P3BI.

Dorée, A.G. (2004) Collusion in the Dutch construction industry: an industrial organization perspective. *Building Research and Information,* **32** (2), 146–156.

Douma, S. & Schreuder, H. (1998) *Economic Approaches to Organizations,* 2nd edn. London: Prentice Hall.

Dunn, J.A., Jr (1999) Transportation: policy-level partnerships and project-based partnerships, *American Behavioral Scientist,* **43**(1), 92–106.

Egan, J. (1998) *Rethinking Construction.* London: Department of the Environment, Transport and the Regions.

Ezulike, E.I., Perry, J.P. & Hawwash, K. (1997) The barriers to entry into the PFI market. *Engineering, Construction and Architectural Management,* **4** (3), 179–193.

Fosler, R.S. & Berger, R.A. (1982) *Public-Private Partnership in American Cities.* Lexington: Lexington Books.

Foss, N.J. & Koch, C.A. (1996) Opportunism, organizational economics and the network approach. *Scandinavian Journal of Management,* **12** (2), 189–205.

Gaffney, D., Pollock, A.M., Price, D. & Shaoul, J. (1999) The politics of the private finance initiative and the new NHS. *British Medical Journal,* **319**, 249–253.

Gansler, J.S. (2003) *Moving Towards Market-based Government: The Changing Role of Government as the Provider.* Washington DC: IBM Endowment for the Business of Government.

Gils, M.R. van (1978) De organisatie van organisaties: aspecten van interorganizationele samenwerking. *M&O,* **1**, 9–31.

Godfroij, A.J.A. (1992) Dynamische netwerken. *M&O,* **4**, 365–375.

Granovetter, M. (1985). Economic action and social structure: the problem of embeddedness. *American Journal of Sociology,* **4** (3), 481–510.

Håkansson, H. (1989) *Corporate Technological Behavior.* London: Routledge.

Håkansson, H. & Johansson, J. (1993) The network as a governance structure: interfirm cooperation beyond markets and hierarchies. In: G. Grabner (Ed) *The Embedded Firm; Understanding Networks: Actors, Resources and Processes in Interfirm Cooperation.* London: Routledge.

Hall, J. (1998) Private opportunity, public benefit? *Fiscal Studies*, **19** (2), 121–140.

Hazeu, C.K. (2000) *Institutionele Economie.* Bussum: Coutinho.

Hellgren, B. & Sternberg, T. (1995) Design and implementation in major investments: a project network approach. *Scandinavian Journal of Management*, **11** (4), 377–394.

Jacobs, B.D. (1998) Bureaupolitics and public/private partnerships in economic development in the British West Midlands. In: N. Walzer & B.D. Jacobs (Eds) *Public–Private Partnerships for Local Economic Development* (pp. 19–46). Westport: Praeger.

Kickert, W.J.M., Klijn, E.H. & Koppenjan, J.F.M. (1999) *Managing Complex Networks: Strategies for the Public Sector.* London: Sage.

Klaver, D. & Siccama, J. (1974) Integratie van politicologie en economie. *Acta Politica*, **9**, 125–161.

Klijn, E.H. & Teisman, G.R. (2003) Institutional and strategic barriers to public–private partnership: an analysis of Dutch cases. *Public Money and Management*, **23** (3), 137–147.

Knowledge Centre Public–Private Partnerships (2004) *Progress Report.* The Hague: Ministry of Finance.

Koppenjan, J. (2005) The formation of public–private partnerships: lessons from nine transport infrastructure projects in the Netherlands. *Public Administration*, **83** (1), 135–157.

Laws, D., Susskind, L., Abrams, J., Anderson, J., Chapman, G., Rubenstein, E. & Vadgama, J. (2001) *Public Entrepreneurship Networks.* Cambridge, MA: Massachusetts Institute of Technology, Department of Urban Studies and Planning.

Lemstra, W. (1996) Samenwerking tussen overheid en bedrijfsleven: utopie of werkelijkheid? *Bedrijfskunde*, **68** (3), 44–50.

Lemstra, W., Kuijken, W.J. & Versteden, C.J.N. (1996) *Handboek Overheidsmanagement.* Alphen aan den Rijn/Brussel: Samsom H.D. Tjeenk Willink.

Li Bing & Akintoye, A. (2003) An overview of public–private partnership. In: A. Akintoye, M. Beck & C. Hardcastle (Eds) *Public–Private Partnerships: Managing Risks and Opportunities* (pp. 3–30). Oxford: Blackwell Science.

Li Bing, Akintoye, A., Edwards, P.J. & Hardcastle, C. (2005) The allocation of risk in PPP/PFI construction projects in the UK. *International Journal of Project Management*, **23**, 25–35.

Linder, S.H. (1999) Coming to terms with the public–private partnership. *American Behavioral Scientist*, **43** (1), 35–51.

Luscuere, C. (1978) Samenwerking tussen organisaties: ideologieën en dilemma's. *M&O*, **1**, 49–60.

Madhok, A. (1995) Opportunism and trust in joint venture relationships: an exploratory study and a model. *Scandinavian Journal of Management*, **11**(1), 57–74.

Ministry of Economic Affairs (1999) *Nota Ruimtelijk Economisch Beleid: Dynamiek in Netwerken.* The Hague: Ministry of Economic Affairs.

National Audit Office (2001) *Managing the Relationship to Secure a Successful Partnership in PFI Projects.* London: National Audit Office.

Nelson, S. (2001) The nature of partnership in urban renewal in Paris and London. *European Planning Studies*, **9** (4), 483–502.

Nijkamp, P., Burch, M. van der & Vindigni, G. (2002) A comparative institutional evaluation of public–private partnerships in Dutch urban land-use and revitalisation projects. *Urban Studies*, **39** (10), 1865–1880.

Packendorff, J. (1995) Inquiring into the temporary organization: new directions for project management research. *Scandinavian Journal of Management*, **11** (4), 319–333.

Peters, B.G. (1997) With a little help from our friends: public–private partnerships as institutions and instruments. In: J. Pierre (Ed) *Partnerships in Urban Governance: European and American Experiences* (pp. 11–33). London: Macmillan.

Pongsiri, N. (2002) Regulation and public–private partnerships. *International Journal of Public Sector Management*, **15** (6), 487–495.

Powell, W.W. (1991) Neither market nor hierarchy: network forms of organization. In: G. Thompson, J. Frances, R. Levačič et al. (Eds) *Markets, Hierarchies and Networks: The Coordination of Social Life* (pp. 265–276). London: Sage.

Raad voor het Openbaar Bestuur (1998) *De Overheid de Markt in- of Uitprijzen*. The Hague: Raad voor het Openbaar Bestuur.

Reijniers, J.J.A.M. (1994) Organization of public–private partnership projects. *International Journal of Project Management*, **12** (3), 137–142.

Rosenau, P.V. (1999) The strengths and weaknesses of public–private policy partnerships: editor's introduction. *American Behavioral Scientist*, **43** (1), 10–34.

Savas, E. (2000) *Privatization and Public–Private Partnerships*. New York: Chatham House.

Schultz, J.F.H. (1994) *EDI: Kansspel, Machtsspel of Samenspel*. Alphen aan den Rijn: Samsom Bedrijfsinformatie.

Spiering, W.D. & Dewulf, G.P.M.R. (2001) *Publiek–Private Samenwerking bij Infrastructurele en Stedelijke Projecten*. Enschede: P3BI.

Stainback, J. (2000) *Public/Private Finance and Development: Methodology, Deal Structuring, Developer Solicitation*. New York: John Wiley & Sons.

Stephenson, M.O. (1991) Whither the public–private partnership. *Urban Affairs Quarterly*, **27** (1), 109–127.

Stoker, G. (1997) Public–private partnerships and urban governance. In: J. Pierre (Ed) *Partnerships in Urban Governance: European and American Experiences*, (pp. 34–51). London: Macmillan.

Teisman, G.R. (1998) *Complexe Besluitvorming*. Maarssen: Elsevier Bedrijfsinformatie.

Trist, E. (1983) Referent organizations and the development of interorganizational domains. *Human Relations*, **36** (3), 269–284.

Turner, J.R. (1993) *The Handbook of Project-Based Management: Improving the Processes for Achieving Strategic Objectives*. London: Henley Management Series.

Van Damme, E. (2001) Marktwerking vereist maatwerk. *Maandschrift Economie*, **65** (3), 185–207.

Wetenschappelijke Raad voor het Regeringsbeleid (2001) *Safeguarding the Public Interest: Summary of the 56th Report*. The Hague: SDU.

Williamson, O.E. (1983) *Markets and Hierarchies: Analysis and Antitrust Implications*. New York: Free Press.

Further Reading

Akintoye, A., Beck, M. & Hardcastle, C. (Eds) (2003) *Public–Private Partnerships: Managing Risks and Opportunities*. Oxford: Blackwell Science.

Beder, H. (1984) *Realizing the Potential of Interorganizational Cooperation*. San Francisco: Jossey-Bass.

Bennett, R.J. & Krebs, G. (1991) *Public–Private Partnership Initiation in Britain and Germany*. London: Belhaven.

Berry, J., McGreal, S. & Deddis, B. (1993) *Urban Regeneration: Property Investment and Development*. London: E&FN Spon.

Börzel, T.A. & Risse, T. (2002) Public–private partnerships: effective and legitimate tools of international governance? In: E. Grande & W. Pauly (Eds) *Complex Sovereignty: on the Reconstitution of Political Authority in the 21st Century* (unpublished).

Bozeman, B. (1987) *All Organizations are Public: Bridging Public and Private Organizational Theories*. San Francisco: Jossey-Bass.

Bongenaar, A. (2001) *Corporate Governance and Public Private Partnership: The Case of Japan*. Utrecht: Nederlandse Geografische Studies.

Estache, A., Romero, M. & Strong, J. (2001). *Privatization and Regulation of Transport Infrastructure: Guidelines for Policymakers and Regulators*. Washington DC: WBI Development Studies.

Fischbacher, M. & Beaumont, P.B. (2003) PFI, public–private partnerships and the neglected importance of process: stakeholders and the employment dimension. *Public Money and Management*, **23** (3), 171–177.

Galaskiewicz, J. & Mizruchi, M.S. (1993) Networks of interorganizational relations. *Sociological Methods and Research*, **22** (1), 46–70.

Hall, R.H. (1991) *Organizations: Structures, Processes and Outcomes*. Englewood Cliffs: Prentice Hall.

Klein Woolthuis, R.J.A. (1999) *Sleeping with the Enemy: Trust, Dependence and Contracts in Interorganizational Relationships*. Enschede: Universiteit Twente.

Knibbe, A. (2002) *Publiek–Private Samenwerking*. Alphen aan den Rijn: Kluwer.

Kolzow, D.R. (1994) Public/private partnership: the economic development organization of the 90s. *Economic Development Review*, **12**, 4–6.

Kouwenhoven, V.P. (1991) *Publiek–Private Samenwerking: Model of Model?* Delft: Eburon.

Kreukels, A.M.J. & Spit, T.J.M. (1990) Public–private partnership in the Netherlands. *Tijdschrift voor Economische en Sociale Geografie*, **5**, 388–392.

Larkin, G.R. (1994) Public–private partnerships in economic development. *Economic Development Review*, **12**, 7–9.

Levitt, R.L. & Kirlin, J.J. (1990) *Managing Development Through Public/Private Negotiations*. Washington DC: Urban Land Institute.

Montanheiro, L. & Spiering, W.D. (Eds) (2001) *Public and Private Sector Partnerships: The Enterprise Governance*. Sheffield: Sheffield Hallam University.

Mulford, C.M. (1984) *Interorganizational Relations*. New York: Human Science Press.

Osborne, S.P. (Ed) (2000) *Public–Private Partnerships: Theory and Practice in International Perspective*. London: Routledge.

Plummer, J. (Ed) (2002) *Focusing Partnerships: A Sourcebook for Municipal Capacity Building in Public–Private Partnerships.* London: Earthscan.

Powell, W.W. (1987) Hybrid organizational arrangements: new form or transitional development. *California Management Review,* **1,** 67–87.

Rijkswaterstaat Steunpunt Opdrachtgeverschap (1997) *Handreiking Bouworganisatievormen.* The Hague: Ministerie van Verkeer & Waterstaat.

Ring, P.S. & Ven, A.H. van de (1994) Developmental processes of cooperative interorganizational relationships. *Academy of Management Review,* **19** (1), 90–118.

Salet, W. & Faludi, A. (Eds) (2000) *The Revival of Strategic Planning.* Amsterdam: KNAW.

Spiering, W.D. (2000) Public–private partnership in city revitalisation. *International Journal of Public–Private Partnerships,* **3** (1), 133–145.

Thompson, G., Frances, J., Levačič, R. et al. (Eds) (1991) *Markets, Hierarchies and Networks: The Coordination of Social Life.* London: Sage.

Vangen, S. & Huxham, C. (2003) Nurturing collaborative relations: building trust in interorganizational collaboration. *Journal of Applied Behavioral Science,* **39** (1), 5–31.

Walzer, N. & York, L. (1998) Public–private partnerships in US cities. In: N. Walzer & B.D. Jacobs (Eds) *Public–Private Partnerships for Local Economic Development* (pp. 47–68). Westport: Praeger.

Weick, K.E. (1979) *The Social Psychology of Organizing.* Reading: Addison-Wesley.

Wilkof, M.V., Wright Brown, D. & Selsky, J.W. (1995) When the stories are different: the influence of corporate culture mismatches on interorganizational relations. *Journal of Applied Science,* **31** (3), 373–388.

3 Procurement Systems in Construction: Europe and USA

Forming coalitions is among the most important construction process activities. The process of selecting the right (private) parties for development, design, construction, maintenance, operation or exploitation, is known as procurement. Public–private partnerships, just like all other kinds of coalitions, have to meet the specific criteria and rules of the procurement systems that are used in a given country.

To define procurement in a straightforward way is difficult, if not impossible. Confusion about concepts and different claims concerning the economic workings and contractual basics of procurement complicate the study of this phenomenon. New developments, insights and reforms are illustrative for procurement, both in practice and in theory. Models are applied in practice and translated into products by which the construction market is organized and the many different actors in the field co-operate (Egan, 1998; National Audit Office, 2001; Dorée, 2004).

From an economic perspective, we can distinguish between PPPs and other forms of procurement systems. Table 3.1 provides a simplified overview of the division of roles and contributions of public and private actors in these different procurement systems: two contractual forms of PPPs are distinguished, concession contracts and joint venture contracts. When using the framework of analysis set up in Chapter 2 (Table 2.3) in practice, important differences are found between these types of PPP so the procurement systems will be described for both.

This chapter aims to present the models of procurement applied in PPPs by giving an overview of the procurement context and an outline of academic and practical discussions. Thus only some basics of procurement in construction in general, and the characteristics of the two forms of PPPs as specific procurement systems are described. Starting points for the current European and American approach to PPPs as procurement systems are used to explain their context.

3.1 Procurement in construction

Winch (2002) distinguishes four main ways for the public sector to procure construction services when positioned in Fig. 3.1:

(1) Maintain an in-house capability: construction services are carried out by specialized agencies within the public sector

Table 3.1   Roles and contributions of public and private actors in four procurement systems in the Netherlands (based on Bult-Spiering, 2003).

Procurement system	Roles and contribution			
	Direction	Procurement process	Risks	Funding
Traditional procurement	Public	Public actor puts one or more works out to tender	Risks and responsibilities for public actor	Costs and revenues for public actor
Innovative procurement	Public	Public actor puts output specified question for overall solution out to tender	Design, build and/ or maintain risks for private actor	Costs and revenues: lump sum for public actor, variable for private actors
PPP: concession contracts	Public	Public actor puts a service question out to tender; rewarded with a concession	Design, build, finance and maintain/operate risks for private actor	Costs and revenues: lump sum for public actor, variable for private actors
PPP: joint venture contracts	Public– private	Joint procurement and shared responsibility	Public–private shared	Cost and revenues: public–private shared
Privatization	Private	Public tasks and competences are transferred to the private sector	Risks and responsibilities for private actor	Costs and revenues for private actor

(2)   Appoint a supplier: selection of a supplier is based on their reputation for having previously completed similar projects

(3)   Launch a concours (competition): the basis is the quality of the solution, rather than its price

(4)   Issue an invitation to tender competitively: the selection process is formalized, and so are the criteria on which the final decision is made.

Within the competitive tender procedure five methods are regularly used (Zhang, 2004a; Li Bing et al., 2005):

(1)   Open competitive tendering
(2)   Invited tendering
(3)   Registered lists
(4)   Project-specific prequalification/shortlisting
(5)   Negotiated tendering.

The most suitable form of procurement depends, according to Winch (2002), on the degree of uncertainty found in the specification. This uncertainty is further subdivided into mission uncertainty and dynamic uncertainty. Large, unique

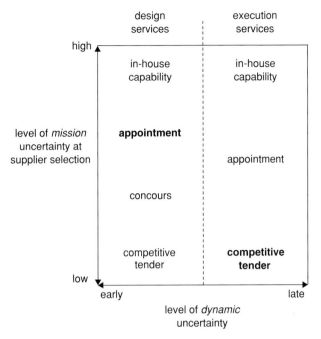

Figure 3.1   Supplier selection methods (Winch, 2002).

projects, for example, have a higher mission uncertainty than small recurring projects, while earlier phases in the project life cycle have a much higher level of dynamic uncertainty. This is shown in Fig. 3.1. The methods in bold type are the most common procurement systems used in each phase.

Although infrastructure projects usually have high levels of both mission and dynamic uncertainty, in-house capability is an unattractive tool for the public sector because the main procurer is responsible for the provision of infrastructure. The projects with most uncertainty are likely to require the appointment of highly specialized resources having relevant experience. Therefore, competitive tendering is the most commonly used means by which suppliers of construction services are selected; in this form of procurement, both the selection process and the selection criteria should be formalized.

Under competitive tendering, prospective suppliers are selected, on the basis of compliance of their offer with the specified tender documents and on the calculated price for supplying the services. The tender documents should provide a detailed description of the construction service to be delivered. Competitive tenders may be open to all bidders, or released only to a pre-established tender list. In the case of open competitions, selection also occurs on the basis of competence criteria when decisions are made on which private organizations are to be invited to tender. In cases of selective competition these matters are addressed in the prequalification process for being invited to tender (Winch, 2002). Table 3.2

Table 3.2 Advantages and disadvantages of competitive tendering (based on Winch, 2002).

Advantages	Disadvantages
Keen price competition encourages production efficiency	Risks caused by limited information on competence of suppliers
Transparency of selection criteria facilitates audit of public sector decisions	High costs of search and selection
Low barriers of entry minimize the risks of cartels forming	Requirement of complete tender documents limits the use of competitive tendering for high uncertainty transactions
	Systematic bias towards underestimated tenders, because errors of omission will win tenders, while errors of inclusion will lose them

provides an overview of the advantages and disadvantages of competitive tendering.

In competitive procurement, private partners organize themselves in alliances, referred to as a joint venture (not to be confused with the joint venture type of PPP). In these private sector joint ventures, risks are internalized and shared among the individual organizations, and the costs of managing the alliance are transaction costs. The private sector problem is to find a partner of appropriate status, wealth and character: Winch (2002) identifies this as the 'Jane Austen problem'.

3.2 PPP procurement systems

Procurement systems are actually organizational structures, formalized in contracts (Dorée, 1996). Different construction sector projects need different contracts to formalize PPPs; these contracts have to be tailor-made because contexts and projects are unique. The two dominant types of PPP procurement systems are concessions and joint ventures. This section describes the main differences between these two types of PPP and they are summarized in Table 3.3.

3.2.1 Concessions

Concession contracting is known variously as private finance initiative (PFI), design–build–finance–maintain (DBFM), design–build–finance–operate (DBFO), build–operate–transfer (BOT), and by many other names (for an overview see Miller, 2000; Winch, 2002; Zhang 2004a). In this book, we concur with Miller (2000) in regarding DBFM, DBFO and BOT as synonyms. These three

Table 3.3 Characteristics of PPP contracts.

	Concession PPP	**Joint venture PPP**
Combination	Integration of construction process activities: technical complexity	Integration of functions: complexity of content
Actors	Central government, provinces, municipalities, consortia/construction companies: vertical organizational complexity	Municipalities, developers, owners: horizontal organizational complexity
Legal structure	Design–build–finance–maintain/operate contract	Separate corporate legal entity
Agreements	Output agreements	Declaration of intent/co-operation agreements

contractual arrangements are delivery methods in which the client procures design, construction, financing, maintenance and operation of an asset and accompanying services as an integrated package delivered by a single contractor. The client provides the initial planning and functional design, while all risks are taken on by the contractor.

Concession PPPs are used for single object developments, especially infrastructure developments. Miller (2000) defines four different segments of infrastructure development:

(1) Capital facilities (buildings, housing, factories and other structures which provide shelter)
(2) Transportation of people, goods and information
(3) Provision of public services and utilities (water, power, waste removal, waste minimization and waste control)
(4) Environmental restoration.

Since we are focusing on the construction sector, in this book infrastructure development comprises capital facilities ('soft' infrastructure) and transportation of people and goods ('hard' infrastructure) (see Section 2.3).

Characteristically for concessions, the public sector buys a service. This service is the availability of a certain product for a certain time, capacity and quality. The service is obtained by granting a concession, where the concessionaire is also responsible for financing the project. Direction is in the hands of the public sector which is the principal or client; risks are transferred to the private sector. In the UK, Spackman (2002) translated the 2000 HM Treasury policy statement as defining the following infrastructure or concession PPP elements:

■ Complete or partial privatization
■ Contracting out with private finance at risk
■ Selling government services in partnership with private sector companies.

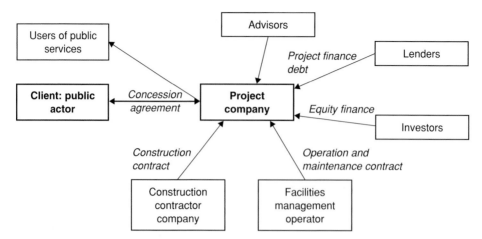

Figure 3.2   Relation of consortium participants to a special purpose company (based on Yescombe, 2002; Asenova et al., 2003; HM Treasury, 2003).

With concession PPPs, the public sector defines what is required to meet public needs and remains the client throughout the life of the contract. The public sector also ensures quality by contract delivery of the outputs it sets. The private sector takes on responsibility for providing a public service, including maintaining and enhancing or constructing the necessary infrastructure (Estache & Serebrisky, 2004).

The concessionaire often is a multidisciplinary consortium specifically formed for a particular PPP project. These consortia consist of organizations that have their core activities in either designing, building, financing, maintaining or operating a certain object. This consortium is the private partner in the concession PPP. The consortium has mainly an executive role. The private consortium is selected through competitive tendering. The private parties participating in the consortium are often organized in a joint special purpose company (SPC): through this vehicle, risks to the consortium members can be limited and the project can be assessed on its own merits. Figure 3.2 shows, in general terms, the relation between an SPC and its participants.

Usually, the consortium has to bear most of the risks involved in the project. Zhang (2004a) distinguishes two categories of risk:

- *Elementary risks*: physical, design, construction, operation and maintenance, technology, finance and revenue-generation
- *Global risks*: political, legal, commercial and environmental.

In most cases all elementary risks are borne by the private partner, whereas global risks are assigned to the PPP participant best able to control them.

Projects that are established through concession PPPs are often of local, regional and national value. Because of this, different levels of government (national, regional and local) are involved on the public side, which all have their own jurisdiction, authorities and responsibilities. These interests cannot be united in an SPC as in the private sector; this causes vertical organizational complexity because of the hierarchical relations between the different levels of government.

Although concession PPPs do not integrate different categories of spatial function (such as real estate, infrastructure and nature), they do integrate construction process activities: concession contracts are based on a life cycle and supply chain management approach. The integration of functions increases the complexity of content of a project, whereas the integration of construction process activities increases technical complexity.

As a result of the integration of activities concession PPPs are formalized in DBFM/O contracts and specified in output agreements. The contract takes up several project phases, ranging from design to detailing through to maintenance and/or operation. Private consortia are often consulted during an earlier project phase, but final selection of the concessionaire only takes place after the development phase (see list below). In many countries today, studies are taking place on how private parties could be involved earlier in the process.

The activities in a PPP life cycle can be subdivided into three phases (Scholten, 2001):

(1) Development
     — Initiation
     — Definition
     — Prequalification
     — Consultation
     — Tender
     — Negotiation
     — Financial close
(2) Construction
(3) Operation and maintenance
     — Operation
     — Transfer.

Tendering procedure for concession PPPs

For concession PPPs, competitive tendering is the most common approach. In the USA this procedure is used, for example, for toll roads constructed under the Intermodal Surface (Transportation Efficiency) Act and for single real estate objects (Stainback, 2000; Zhang, 2004a; Li Bing et al., 2005). According to European Union guidelines, for all projects exceeding the value-added tax (VAT) exclusive value of €5.923 million, competitive tendering procedures must be followed (www.europa.eu.int; accessed June 2005).

The stages in the competitive tendering (also called soliciting) process are not restricted to the actual tender phase, but cover the entire development phase, and therefore mostly include the following (Savas, 2000; Stainback, 2000; Scholten, 2001; Zhang, 2004a):

(1) Market consultation or request for information
(2) Request for (pre)qualification and (pre)qualification
(3) Invitation to tender or request for proposal
(4) Tender evaluation and short-listing
(5) Negotiation with short-listed tenderers
(6) Selection of the tender, award of concession and financial close.

Because of the high costs of tendering for concession infrastructure projects (due to technical, contractual and financial complexity), the prequalification process aims to select a short list (five to eight) of qualified consortia, of which three or four at most are invited to tender after assessment and negotiation. The tender stages will now be discussed separately.

Market consultation or request for information (RFI)
This request is issued to solicit ideas and concepts from the private sector before a request for qualification (RFQ) is issued. In addition, the RFI is sometimes used to corroborate whether the correct issues have been addressed by the public sector to develop the project, and could function as an announcement of a project opportunity.

Request for (pre)qualification (RFQ) and (pre)qualification
The RFQ process comprises the following stages:

- Develop the RFQ
- Review and approve the RFQ
- Set up and document the evaluation criteria used to select a developer
- Identify the private consortia to receive the RFQ
- Produce and issue the RFQ
- Give the consortia the prearranged time to prepare their proposals
- Complete a preproposal conference or presentation
- Evaluate the consortia's proposals
- Review the results of evaluation
- Announce the short-listed development teams.

Different prequalification methods can be used, for example the binary method (select consortia that meet all the pre-established basic criteria and reject those consortia that do not meet these criteria), the simple scoring method, multi-attribute analysis, and any of these three methods in combination with an outline tender proposal (see Zhang, 2004b for further details).

The evaluation criteria for the RFQ stage are often classified in five categories:

(1) *Information on the consortia* General and financial reports on the proposed concessionaires, i.e. the private consortia, and the consortia's management structure

(2) *Information and qualifications of the consortia members* Past work carried out by consortia, existing workload of the consortia, experience of the consortia's legal and financial advisers in PPP projects, consortia's experience in managing projects, health and safety records, quality assurance systems and industrial relations

(3) *Consortia's initial vision for the project* Range of technical, operational, and financial capabilities necessary for the project, with other resources and references

(4) *Initial financial concepts* Proposed sources of finance and debt/equity ratio

(5) *Relationship* Anticipated working relationship between the public and private partners.

Invitation to tender or request for proposal (RFP)

This invitation takes place through the project advertisement. In Europe, advertisements are published in the Official Journal of the European Community (OJEC). Advertisements must be drafted in broad output terms to achieve maximum flexibility for potential bidders and to exploit strategic opportunities.

The purpose of the RFP is to get a highly detailed and technical response from the private sector, using the requirements and format drafted by the public partner. Among the many issues that need to be addressed by the consortia are the finance plan and model, deal structure, master plan, phasing, budget and the planned interaction with the public partner. Required documents for this phase are the project brief, instructions to tender, the draft concession agreement, a financial analysis model and an outline of the tender evaluation method.

Tender evaluation and short-listing

Zhang (2004b) distinguishes several methods for evaluating competitive bids:

- Net present value (NPV) method
- Multi-attribute analysis
- Kepner–Tregoe decision analysis technique
- Two-envelope method
- NPV method plus scoring method
- Binary method plus NPV method
- Simple scoring method.

Negotiation with short-listed tenderers

After assessment of the initial tender proposals, the client may negotiate to achieve better value. Bidders may be required to submit revised proposals and are asked to make their best and final offer (BAFO).

Selection of the tender, award of concession and financial close

Financial close, that is the conclusive financing arrangement, is reached at the end of the negotiation phase. After selection of the consortium or concessionaire, the financiers' advisers assess the project on legislation, tax law and technical feasibility issues. The comments and requirements of the financiers will affect the commercial deal between the public sector and the consortium. After the financial close, the consortium can start detailing its design to start construction.

The private consortium selection should be based on the best value approach: the concessionaires that are asked for a BAFO, are selected against multiple evaluation criteria. Selection, though, is still merely based on the economically most profitable solution, expressed in price-related criteria. The lowest bid price is usually the most important or even the sole selection criterion by which a contractor is selected (Dewulf et al., 2004; Zhang, 2004a).

3.2.2 Joint ventures

Joint-venture PPPs are mostly adopted in urban area development projects. Joint commissioning and responsibilities are characteristics of these PPPs. Direction is in both public and private hands, and public and private partners both provide resources. Risks, revenues and losses are shared.

This type of partnership is often referred to as a 'true' PPP (Bennett et al., 2000; Li Bing & Akintoye, 2003). Public and private actors can bundle their resources and generate a mutual return; both are active shareholders in a joint operating company (Li Bing & Akintoye, 2003).

Public and private sector goals do not necessarily have to be identical for a joint venture PPP to succeed, but the aims have to be compatible and lead to a shared outcome (Bennett et al., 2000). For example, the private partner may be interested in commercial returns on investments, while the public partner tries to establish social revenues, yet both share the common goal of creating a viable and sustainable solution for a social problem in an urban area.

Similar to concessions, joint ventures require a multi-actor approach. In joint ventures, however, the main actors involved in PPPs are local governments, developers and land or property owners; joint-venture PPPs therefore do not operate at a regional or national level.

Under joint-venture PPPs, the public actor is both the regulator and a shareholder in the joint operating company. From this position, it will share in the operating company's profits and help ensure the wider political acceptability of its efforts to safeguard public needs and interests. Although the private actor often has the primary responsibility for performing daily management operations, the public sector continues to play a role at the corporate governance and day-to-day management levels (Bennett et al., 2000; Bult-Spiering, 2003).

Developers can be independent, however; for example in The Netherlands, developers linked to construction companies, financial institutions, housing agencies, as well as investors are important market players. This means that they can have different roles and interests in the construction process. The primary role of the private actors is merely developing, building and/or exploiting assets or projects. Projects that are established through joint venture PPPs are of local importance, which implies that users of the future facilities are quite easy to identify. Joint venture PPPs often concern urban redevelopments. In this case land is often split up, and private property or landowners are powerful stakeholders. This situation causes horizontal organizational complexity, because of the absence of hierarchy between the different stakeholders.

Joint venture PPPs are separate legal entities, with autonomous tasks and responsibilities. To set up such an entity, an initial agreement (often a declaration of intent) should be formulated and, after a positive feasibility study, translated into a co-operation agreement. These PPPs often concern all project phases, from initiative through to exploitation and management. Occasionally, the joint venture starts at a further stage, or ends just before or after the realization of the project.

The private sector is, in joint venture PPPs, invited to develop plans through competition, and selection of the private sector partner is based on project vision and references. The selection of a private partner with whom the government wishes to enter the joint venture PPPs, is not subject to public procurement rules. However, jurisprudence of, for example, the European Court of Justice (2001) shows that any action of a government by which it establishes conditions for economic activities is governed by the general principles of the Treaty of Rome 1957. The way the joint venture is constituted must be compatible with the law of the European Union. For example, land ownership is therefore not an exception for exclusive co-operation and exclusion of competition. In the USA, the exclusiveness of private property owners and private property rights are not as highly valued and protected as in former years: 'Eminent domain', embedded in federal and many state laws, is an instrument for dispossessing private owners and making their property available for public use. This instrument is to an increasing extent being used by US local governments to make large-scale developments possible, the attraction being the increasing property tax incomes resulting from them. The case of Kelo versus the City of New London, Connecticut, is illustrative of the on-going discussion on this theme (www.ij.org/pdf_folder/private_property/kelo/kelo-USSC-opinion-6-05.pdf; en.wikipedia.org/wiki/Kelo_v._New_London#The_decision; accessed June 2005).

3.2.3 Comparison

Practice shows that the motives for a partnership strongly determine the further development of the co-operation process. As stated in the ideal model PPP (Section 2.2), the extent to which public and private partners are interdependent is an important factor for success. In urban governance projects this interdependence is strong and it is not surprising that evaluations show these PPP projects as being most successful (Osborne, 2000; Spiering & Dewulf 2001; Nijkamp et al., 2002).

Next to interdependence, the way in which the added value can be defined plays an important role. Several national construction reform initiatives show the relevance of an adequate price/quality ratio and the appointment of value. Again, the value of the partnership can be proved to be better in urban development projects than in infrastructure projects. Since improving real estate value is the dominant driver for the private sector in urban development, investments in quality will contribute to its goal and will therefore result in added value in project content. In infrastructure projects, however, the private sector has a stronger interest in the construction of the objects, and therefore traditionally is less focused on object quality; there is a more dominant focus on financial added value in dividing risks.

The early involvement of private actors facilitates achievement of the optimum possibilities for exchange and added value (Kinnock, 1998; Bennett et al., 2000; Bult-Spiering, 2003). Through early involvement of the private sector, it is assumed that the project can deliver more added value to all parties. For example, in the Netherlands early involvement is easier to achieve in urban area development than in the development of infrastructure: in urban developments, the actual partnership starts earlier (in the initiation phase), while in infrastructure projects the final selection is made after the development phase.

If a PPP covers several project phases, it creates more opportunities for process and financial added value. In that case, actors jointly can find solutions for societal problems, with both social and business economic revenues. The eventual choice for a partner is based on criteria that are ideally derived from goals and starting points of the separate organizations. In addition, added value of the concession PPP should be expressed in innovation and value for money of product and process, the life cycle approach, the transfer of risks, and reductions in cost and time (NAO, 2001; Bult-Spiering, 2003). In practice, however, this added value is not always created (see Chapter 4).

Added value in process is easier to establish in urban projects, because of the limited public sector parties involved. The vertical organizational complexity in infrastructure projects caused by the large number of public sector actors, often leads to inadequate alignment of the public–public co-operation. This problem is less dominant in urban development.

A major disadvantage of joint venture PPPs, however, is the possible conflict of interests for the public sector, because the two different roles as a shareholder on the one hand, and a guardian of public interest on the other, can be contradictory (Knibbe, 2002; Bult-Spiering, 2003).

3.3 European and American context

Although the enthusiasm for private sector involvement depends strongly, among other things, on the current national or state policy, the influence of community or federal policy on PPPs is significant, as exemplified by the London docklands development in the UK.[1]

---

[1] Before 1979 when the Labour Party was in power, the Docklands Joint Committee (DJC) was responsible for remaking London docklands; this committee, under public

3.3.1 Europe

Although all EU member states have an interest in PPPs, their experience of PPP procurement is limited. The UK has the longest and most extensive experience of PPPs in several different sectors. Table 3.4 presents an overview of the experience with PPPs amongst EU members, new EU member states and applicant countries.

Belgium, France, Germany, Greece, Ireland, Portugal and Spain have comprehensive PPP legislation. PPP units have been established at central government level; one of their goals is to promote PPP. This is the case in Austria, Ireland, The Netherlands and the UK, for example (PricewaterhouseCoopers, 2004).

The term public–private partnership is not accurately defined at European Community level. From an EU perspective, the term refers to forms of co-operation between public authorities and business that aim to ensure the funding, construction, renovation, management or maintenance of an infrastructure or the provision of a service. Member states consider PPPs to be adequate structures for undertaking infrastructure projects, such as in transport, public health, education and national security. PPPs offer possibilities for breathing new life into the trans-European transport networks (TENs), development of which had fallen behind because of a lack of funding (Commission of the European Communities, 2004a).

European Community law does not lay down specific rules for PPP arrangements, but contracts with a third party are subject to rules and principles resulting from the Treaty of Rome 1957: the principles of equal treatment, mutual recognition, proportionality and transparency. These principles are translated in the directives for the co-ordination of procedures for the award of public contracts (Commission of the European Communities, 2004b). The transparency principle, for example, safeguards the transparency of the contracting public authority's steps in all phases of the purchasing procedure. Hence, the premise on which the rules for public procurement are based is that government contracts should be awarded through a transparent and verifiable public procurement procedure under which all candidates are treated equally. The government has a public responsibility and spends public funds. From this point of view, the government must award contracts to develop public projects through a process of competitive tendering. This also applies, in principle, where the development of a project is to take place in a PPP, even if the initiative for the partnership has come from the market. Unsolicited proposals are not possible in Europe due to EU procurement directives.

---

management, had a minority of private sector representatives. When in 1979 the Conservative Party came to power, the DJC was replaced by the London Docklands Development Corporation (LDDC), an expediter for private investment and market-based development, in accordance with the US model.

Table 3.4 Summary of PPPs by country and sector. Reproduced with permission from PricewaterhouseCoopers (2004) *Developing Public Private Partnerships in New Europe.* PricewaterhouseCoopers, London.

	Central accommodation	Airports	Defence	Housing	Health & hospitals	IT	Ports	Prisons	Heavy railway	Light railway	Roads	Schools	Sports & leisure	Water & waste water (incl solid waste)
Member states														
Austria	O	O◐			◐	O		O	◐O◐		◐O◐◐	O O◐◐		O◐
Belgium	O		O							O	O◐	O O	◐	O
Denmark	◐◐◑◐O	◐O◑	O O O●	◐	◐◑◑O		◐	O	◐O◐	●●	◐◑O◐●	◐O O●		O●● ●
Finland	O O	◐O				◐	◐	◐◐	◐◐	◐●	●◑◑◑◑	O◐ O●	◐	◐◑
France														
Germany		◐O		◐O	◐◑		O	O O O			◐◑◑● ◐	●	◐ O	◐◐◐
Greece						◐	O		● O O	◐◑				
Ireland	O O	O O	O O	O O	◐◑◑◑◑O●	◐	O			◐◑◑●	●◑◑●O●	◐◑O O	O	◐◐
Italy							O●						●	●
Luxembourg														
Netherlands	O	●	◐●	O	O O◐	●	◐	●	O	O O O O	◐◑O O◐O	O O●	O	◐◐ ●
Norway (not EU)		O◐	◐					◐					◐	O
Portugal	●	●		O	◐						●O			
Spain				O O		O O				O	◐O		O	O O
Sweden														
UK		O O		O O			O				O O	O O●		O O●
New member states														
Cyprus	O							O		O	O◐			◐◐◐
Czech Republic			◐	O	◐						◐◑●◐●	◐		
Estonia				O O										
Hungary														
Latvia														
Lithuania														
Malta	O													
Poland									O	O				
Slovakia					◐								◐	
Slovenia				O										
Applicant countries														
Bulgaria		O												◐◐●
Romania		●							O	O	O◐◐			
Turkey											O			

Key: O, discussions ongoing; ◐, projects in procurement; ◑, many procured projects; ◒, many procured projects, some projects closed; ●, substantial number of closed projects; ⬤, substantial number of closed projects, majority of them in operation.

The EU procurement system has been criticized, mainly because of its formal nature and restricted applicability. Many member governments already held long term relationships with the private sector and the EU procurement directives forced them to open their tendering procedures (Teisman & Klijn, 2000). A discrepancy still exists between the EU procurement directives and PPP practice. Braun (2003) mentions two important problems caused by this difference:

(1)   *Inflexibility of specifications* The rules do not allow a substantial change in the project, which significantly reduces the space for innovative inputs from bidders throughout the procurement process
(2)   *Uncertainty on the use of the negotiated procedure* In practice neither the restricted nor the open procedure has proven adequate to secure workable solutions in PPP projects; the rules still do not provide an opportunity to use the negotiated procedure on a regular basis.

Although the EU directives do not make PPPs easier to apply in the separate member states, the need for co-operation, partnerships and interaction amongst organizations is still growing because of the increasing overlap between public and private sector activities (see Chapters 1 and 2). New directives that attempted to give solutions for the problems listed above, have been published (Commission of the European Communities, 2004a,b). Under the EU procurement directives, three procedures can be used (Eur-Lex, 2004):

(1)   Open procedures, in which any interested economic operator may submit a tender
(2)   Restricted procedures, in which any economic operator may request to participate and only candidates invited by the contracting entity may submit a tender
(3)   Competitive dialogue, in which any economic operator may request to participate and where the contracting authority conducts a dialogue with the candidates admitted to that procedure. The aim of this procedure is to develop one or more suitable alternatives capable of meeting its requirements and on the basis of which candidates chosen are invited to tender.

Together with these new directives, the Commission issued a Green Paper on PPPs, the contents of which are described in the next section.

EU Green Paper on PPPs

The Green Paper entitled Public–private partnerships and community law on public contracts and concessions (Commission of the European Communities, 2004c), concerns one of the priorities identified by the Commission in its internal market strategy for 2003–2006; it also contributes to the measures planned as part of the initiative on growth in Europe and was published to launch a debate on the best way to ensure that PPPs develop in a context of effective competition and legal clarity with the aim of exploring how procurement law applies to the different forms of PPP developing in the member states. The Green Paper assesses whether there is a need to clarify, complement or improve the current legal framework at the European level.

The Green Paper is meant to analyse the phenomenon of PPPs with regard to Community law on public procurement and concessions. Under Community law there is no specific system governing PPPs: PPPs created for contracts, which qualify as 'public contracts' under the EU procurement directives co-ordinating procedures for the award of public contracts, must comply with the detailed provisions of those directives; however, 'works concessions' are covered by only a few scattered provisions of secondary legislation and 'service concessions' are not covered by the 'public contracts' directives at all. Nevertheless, all contracts in which a public body awards work involving an economic activity to a third party, whether covered by secondary legislation or not, must be examined in the light of the rules and principles of the Treaty of Rome 1957. This applies particularly for Articles 43 to 49 on the freedom of establishment and the freedom to supply services. These principles include, in particular, the principles of transparency, equal treatment, proportionality and mutual recognition. The Green Paper therefore describes the ways in which the rules and the principles deriving from Community law on public contracts and concessions are applied when a private partner is being selected in the context of different types of PPP, for the duration of the contract. A set of questions intended to find out more about how these rules and principles work in practice are posed to enable the Commission to determine whether they are sufficiently clear and suitable for the requirements and characteristics of PPPs.

The EU rules governing the choice of a private partner, however, have been co-ordinated in the Community at various levels and to varying extents, so a wide variety of approaches are still possible at national level. The Green Paper thus addresses various topics:

- The framework for the procedures for selecting a private partner (competitive dialogue procedure for certain PPP operations qualifying as public contracts, minimal framework in secondary legislation for works concessions, no framework in secondary legislation for service concessions)
- Privately initiated PPPs
- Contractual framework and contract amendments during the life of a PPP
- Subcontracting.

Public–private partnerships created on the basis of purely contractual links ('contractual PPPs' or concessions) and PPPs involving joint participation of a public partner and a private partner in a mixed capital legal entity ('institutional PPPs' or joint ventures) are both addressed in the Green Paper which also defines the concept of concession in Community law and the obligations for public authorities when selecting the concessionaires (Commission of the European Communities, 2004c).

Concession PPPs according to the EU

In these PPPs the partnership between the public and the private sector is based solely on contractual links. One or more tasks are assigned to the private partner: these can include design, funding, execution, renovation or exploitation of a work or service. In the concessive model, there is a direct link between the private partner and the final user: the private partner provides a public service under control of the public partner. The method of remuneration is charging the users of the services or the remuneration takes the form of regular payments by the public partner.

Joint venture PPPs according to the EU

These PPPs involve co-operation between the public and private sector within a distinct entity. This entity is held jointly by the public and the private partner and has the task of ensuring the delivery of a work or service for the benefit of the public. The law on public contracts and concessions does not apply to these mixed-capital entities, but the general principles of the Treaty still are effective. In selecting the private partner, characteristics of its offer (economically most advantageous) should therefore be considered, next to criteria based on capital contribution and experience. If the mixed entity has the quality of a contracting body, it should comply with the law applicable to public contracts and concessions when awarding tasks. In other words, the private partner should not profit from its privileged position in the mixed entity to reserve for itself certain tasks without a prior call for competition.

Public consultation results

A public consultation on the questions raised in the Green Paper was launched by the Commission in April 2004, to which the European Economic and Social Committee and the Committee of the Regions responded. The general conclusions of the resulting report (Commission of the European Communities, 2005) concern the horizontal PPP initiative, selection of the private partner and findings on concession and joint venture PPPs. Thus, a horizontal PPP instrument could cover generally applicable procedural rules, a definition of PPPs and general principles for invitations to tender. A slight majority of contributors clearly oppose horizontal PPP initiatives at EC level. On the subject of selection, most stakeholders appreciate Community initiatives in concessions; for example, through clarifying definitions and core principles of the award procedure. Although the majority of stakeholders believe the current advertisement procedure is sufficient to attract non-national operators, most stakeholders argue for some sort of encouragement for private initiative PPPs. However, only a minority of contributors support an EC initiative on the contractual framework for PPPs. There is no agreement on whether the Community law on public contracts and concessions is complied with by joint public–private entities. Most contributors favour an EC initiative on clarifying applicable Community rules.

On the basis of these results, a communication indicating the Commission's preferences will be published. Possible measures to increase fair competition include legislation, interpretative communications, initiatives to improve the co-ordination of national practice and exchange of good practice between member states.

For the EU, it is currently crucial to complete the trans-European transport networks (TENs) that can interconnect and harmonize national transport networks. Decongested and sustainable mobility requires action at (inter)national, regional and local levels; PPPs can contribute to these policy tasks.

As a European Commission member, Kinnock (1998) poses five conditions for successful PPPs in the EU:

(1) Public–private collaboration should start as early as possible in the life cycle of each particular project
(2) The public sector must define its aims in terms of the nature of the improvement to be gained, as early as possible in the life cycle of the project
(3) The best framework for developing a project is an autonomous, ad hoc company, with the sole purpose of implementing a project
(4) Public–private partnerships must be considered for both TEN priority projects and as a means of promoting smaller infrastructure developments
(5) Each actor in the PPP should bear the risks it is best able to control and has the most interest in controlling

---

3.3.2 USA

Historical evolution of public procurement policies in the USA shows a wide variety of procurement strategies, in which different delivery methods are combined. Methods range from design–build–operate (DBO) and DBFO to separate procurement of these services as in design–bid–build (DBB) and design–build (DB) (Pietroforte & Miller, 2002). Research shows that two phases in the history of procurement in the USA can be distinguished (Miller, 2000):

(1) *1789 to pre-Depression* The use of combined delivery methods, of which 60% were funded by the private sector. Indirect finance was used for most canals, commercial docks, post roads, railroads, telegraph, telephone and power. The most commonly used structures resemble DBO and BOT contracts.

(2) *Post World War II to 2000* Segmented procurement of design and construction services, and direct government funding of public projects. This approach started with the economic policy shift of the Roosevelt administration.

In the 1990s, the market for DB construction grew dramatically in the public sector. The market used to be dominated by large turnkey contractors, but new general contractors with in-house or joint-venture design capabilities did enter. On a smaller scale, US public agencies have been experimenting (again) with combined delivery methods since the 1980s, mainly in transportation and (waste) water projects. However, the results of the revaluation of DBO and BOT projects have been mixed. High bid and proposal costs persuaded the public sector to limit competition or to accept unsolicited proposals. Different proposal requirements, disturbing government signals and vague requests for proposals impeded learning curves and economies of scale. Private sector transaction costs increased and therefore the interest of this sector lessened (Pietroforte & Miller, 2002).

The 'US way' of procurement has its base in three value streams (Anheier & Moulton, 2000):

■ Individual freedom, formal equity before the law, and due process
■ High levels of tolerance for significant disparities in material wealth and well-being combined with a belief in individual advancement and responsibility ('the American Dream')
■ A 'taken for granted-ness' of the US government and best design for the political constitution of society and system of government that requires only fine-tuning and never major overhauls to maintain and perfect it.

Miller (2000) developed the quadrant model to summarize the project delivery methods for infrastructure in the USA; this model is presented in Fig. 3.3. Integration of activities in the construction process increases on moving from left to right on the horizontal axis. The delivery methods DBOT and DBOM, for example, in quadrants II and IV cover all construction process activities. Quadrants I and IV show the project delivery methods where the public sector pushes specific projects directly through current cash assumption. Quadrants II and III show methods in which the public sector stimulates specific projects indirectly through incentives that are supposed to encourage the private sector to participate in these projects. The delivery methods in quadrant II are similar to the methods in quadrant I, except for the cash flow. In quadrant II, cash flow to support design, construction, maintenance and operations is generated solely from the private investor or the project itself; in quadrant I, the government provides the project's funding.

The National Council for Public–Private Partnerships (NCPPP) distinguishes a broad range of PPP contracts that apply in the USA, as does the General Accounting Office (1999). All the following are considered to be PPPs: BOT, build–transfer–operate (BTO), build–own–operate (BOO), buy–build–operate

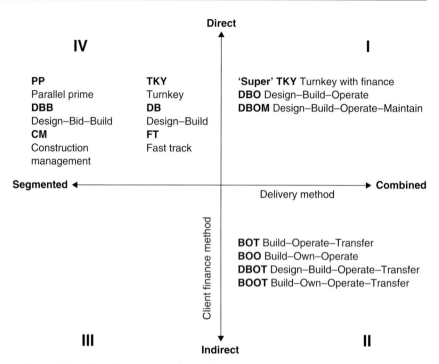

Figure 3.3   Classification of project delivery methods according to the four quadrants (Miller, 2000; Pietroforte & Miller, 2002). Reproduced with permission of Springer-Verlag (Kluwer Academic Publishers), Boston.

(BBO), operations and maintenance, operations, maintenance and management, DB, design–build–maintain (DBM), DBO, developer finance, enhanced use leasing (EUL), lease–develop–operate (LDO) or build–develop–operate (BDO), lease–purchase, sale–leaseback, tax-exempt lease, and turnkey contracts. However, in this book we use a narrower definition of PPPs and distinguish between concession and joint venture PPPs. DBFM, DBFO and BOT are considered to be concession-type contracts.

Just as in European countries, there has been no such thing as a standard project delivery method in American history. The DBFO delivery method was widely used for public infrastructure in the USA after 1789, but has not been commonly applied since World War II (Miller, 2000). Examples of early DBFO projects are the Illinois Central Railroad (1850), the Brooklyn Bridge (1883) and the New York Subway Contract 1 (1940).

In the USA, as in most developed countries, limited direct funding possibilities for growing infrastructure needs have caused a shift away from the delivery methods based on public funding (Pietroforte & Miller, 2002). The 'new' strategy is increasingly based on public–private co-operation in a structured and competitive environment. Starting points are the concurrent use of alternative project delivery and finance methods, a continuous programme of criteria assessment, a good understanding of financial structures and scenario approaches to support decision making in practice (Miller, 2000).

The Federal Acquisition Regulation (FAR) is the primary regulation for use by all federal executive agencies in their acquisitions of supplies and services with appropriate funds. It became effective in 1984 and is issued in applicable laws under the joint authorities of the Administrator of General Services, the Secretary of Defense, and the Administrator for the National Aeronautics and Space Administration, under the broad guidelines of the Administrator, Office of Federal Procurement Policy, Office of Management and Budget. The FAR precluded agency acquisition regulations to those necessary to implement FAR policies and procedures within an agency, and provides for co-ordination, simplicity and uniformity in the Federal Acquisition Process. The Federal Acquisition System is established for the codification and publication of uniform policies and procedures by all executive agencies (General Service Administration et al., 2005).

The solicitation of bids is described in subpart 14.2 of the FAR (part 14 is about sealed bidding); specific rules for construction are described in part 36 of the FAR. Next to traditional contract procurement, part 36 describes the procedure for DB contracts. The FAR does not explicitly refer to the procurement of DBFM/O or BOT contracts. A separate part of the FAR is on the general policy for handling unsolicited proposals (subpart 15.6 under part 15 about Contracting by Negotiation). Because it is the policy of the US Government to encourage submission of new, innovative ideas, these ideas may be submitted as unsolicited proposals when they do not fall under topic areas publicized under government programmes or techniques.

Federal departments have their own guidelines under the FAR that specify procurement procedures for their specific products. These guidelines, in many cases, also go into unsolicited proposals.

The general model for procurement of infrastructure facilities at state level is laid down in the US State and local American Bar Association (ABA) *Model Procurement Code*. This Code was reprinted in 2000 and brings the 1979 edition of the Code to the forefront of leadership in good procurement practices (American Bar Association, 2000). It provides procedures for electronic communications, extends the opportunities for co-operative buying among state and local governments, adds flexibility to procurement methods, and adds DB, DBO and DBFO to the menu for delivery options. The purposes of the Code are as follows:

(1)   Simplify, clarify, and modernize the law governing procurement by the specified state
(2)   Permit the continued development of procurement polices and practices
(3)   Make as consistent as possible the procurement laws among the various jurisdictions
(4)   Provide for increased public confidence in the procedures followed in public procurement

(5) Ensure the fair and equitable treatment of all persons who deal with the procurement system of the specified state

(6) Provide increased economy in the specified state's procurement activities and to maximize to the fullest extent practicable the purchasing value of public funds of the specified state

(7) Foster effective board-based competition within the free enterprise system

(8) Provide safeguards for the maintenance of a procurement system of quality and integrity

(9) Obtain in a cost-effective and responsive manner the materials, services and construction required by the specified state's agencies in order for those agencies to better serve this specified state's business and residents.

Next to this, the Code contains passages on the applicability of the general principles of law and on the requirement of good faith. The Code states that, in principle, contracts for DB, DBOM or DBFO/M project delivery must use competitive sealed proposals as the method of source selection.

Despite the general US policy, and the general procurement code at state level, all US states determine their own legislation and regulation. Consequently, the procurement process is also regulated per state, so individual states prescribe how to submit a bid and have their own policy in allowing unsolicited proposals. Over 15 states have to date enacted some form of legislation enabling PPPs, although with a very broad definition of PPP. For example, the State of Georgia allows unsolicited public–private proposals under its legislation. The Official Code of Georgia provides a framework providing general powers to the Department of Transport to accept unsolicited public–private proposals and advertise for competing proposals, provides authority to evaluate the unsolicited or competing proposal and provides the Department of Transport with authority to contract for public–private transportation initiatives.

The decision to start with a PPP project is merely politically inspired and the initiative for concessions is public in most cases. In addition, the budget is a determinative factor to start a project: money must be available and the considered project needs to have priority over other kinds of projects. The decision-making process concerning concession projects, takes place at state, district and local level. Sometimes, depending on the nature of the project, issues may become federal political issues.

3.4 Summary

Procurement in construction is a wide-ranging and complex subject. In relation to PPPs, two procurement systems can be distinguished: concessions and joint ventures. In this chapter, we have described some basic principles of procurement in construction and the central characteristics of procurement procedures in concession PPPs and in joint venture PPPs. We have also examined the context for PPP procurement in Europe and in the USA. General principles on which the rules for procurement are based both in Europe and in the USA are clarity and equal treatment.

Chapter 4 further analyses concession PPPs; these are illustrated with examples in Chapter 5; Chapters 6 and 7 examine joint venture PPPs.

References

American Bar Association House of Delegates (2000) *The Model Procurement Code Revision Project: Excerpts.* Chicago: American Bar Association.

Anheier, H.K. & Moulton, L. (2000) Public–private partnerships in the United States: historical patterns and current trends. In: S.P. Osborne (Ed) *Public–Private Partnerships: Theory and Practice in International Perspective* (pp. 103–120). London: Routledge.

Asenova, D., Beck, M., Ackintoye, A., Hardcastle, C. & Chiyio, E. (2003) Obstacles to best value in NHS PFI projects. *Journal for Finance and Management in Public Services*, **4** (1), 33–49.

Bennett, E., James, S. & Grohmann, P. (2000) *Joint Venture Public–Private Partnerships for Urban Environmental Services.* New York: Public Private Partnerships for the Urban Environment.

Braun, P. (2003) Strict compliance versus commercial reality: the practical application of EC procurement law to the UK's private finance initiative. *European Law Journal*, **9** (5), 575–598.

Bult-Spiering, M. (2003) *Publiek–Private Samenwerking: De Interactie Centraal.* Utrecht: Lemma.

Commission of the European Communities (2004a) Directive 2004/17/EC of the European Parliament and of the Council of 31 March 2004 co-ordinating the procurement procedures of entities operating in the water, energy, transport and postal services sectors. Brussels: European Union.

Commission of the European Communities (2004b) Directive 2004/18/EC of the European Parliament and of the Council of 31 March 2004 on the co-ordination of procedures for the award of public works contracts, public supply contracts and public service contracts. Brussels: European Union.

Commission of the European Communities (2004c) *Green Paper on Public–Private Partnerships and Community Law on Public Contracts and Concessions.* Brussels: European Union.

Commission of the European Communities (2005) *Report on the Public Consultation on the Green Paper on Pubic–Private Partnerships and Community Law on Public Contracts and Concessions.* Brussels: European Union.

Dewulf, G., Bult-Spiering, M. & Blanken, A. (2004) *Opportunities for PFI in The Netherlands.* Enschede: P3BI.

Dorée, A.G. (1996) *Gemeentelijk Aanbesteden.* Enschede: University of Twente.

Dorée, A.G. (2004) Collusion in the Dutch construction industry: an industrial organization perspective. *Building Research and Information*, **32** (2), 146–156.

Egan, J. (1998) *Rethinking Construction*. London: Department of the Environment, Transport and the Regions.

Estache, A. & Serebrisky, T. (2004) Where do we stand on transport infrastructure deregulation and public–private partnership? *World Bank Policy Research Working Paper*, **3356**, 1–27.

Eur-Lex (2004) Legislation L 134. *Official Journal of the European Union*, **9**, 1–127.

European Court of Justice (2001) Scala Case C-399/98, judgement of the Court of Justice (Sixth Chamber) of 12 July 2001.

General Accounting Office (1999) *Public–Private Partnerships: Terms Related to Building and Facility Partnerships*. Washington DC: General Accounting Office.

General Service Administration (GSA), Department of Defense (DoD) & National Aeronautics and Space Administration (NASA) (2005) *Federal Acquisition Regulation*. Washington DC: GSA, DoD, NASA.

HM Treasury (2003) *PFI: Meeting the Investment Challenge*. London: HM Treasury.

Kinnock, N. (1998) Transport policy needs at the turn of the century. *European Business Journal*, **10** (3), 122–129.

Li Bing, Akintoye, A., Edwards, P.J. & Hardcastle, C. (2005) The allocation of risk in PPP/PFI construction projects in the UK. *International Journal of Project Management*, **23**, 25–35.

Miller, J.B. (2000) *Principles of Public and Private Infrastructure Delivery*. Boston: Kluwer Academic.

National Audit Office (2001) Managing the relationship to secure a successful partnership in PFI projects. London: National Audit Office.

Osborne, S.P. (Ed) (2000) *Public–Private Partnerships: Theory and Practice in International Perspective*. London: Routledge.

Pietroforte, R. & Miller, J.B. (2002) Procurement methods for US infrastructure: historical perspectives and recent trends. *Building Research and Information*, **30** (6), 425–434.

PricewaterhouseCoopers (2004) *Developing Public Private Partnerships in New Europe*. Europe: PricewaterhouseCoopers.

Savas, E. (2000) *Privatization and Public–Private Partnerships*. New York: Chatham House.

Scholten, E.G. (2001) *The monitoring of risks in project financed integrated projects*. Master's Thesis, University of Twente. Maarssen: Strukton.

Spackman, M. (2002) Public–private partnerships: lessons from the British approach. *Economic Systems*, **26**, 283–301.

Spiering, W.D. & Dewulf, G.P.M.R. (2001) *Publiek–Private Samenwerking bij Infrastructurele en Stedelijke Projecten*. Enschede: P3BI.

Stainback, J. (2000) *Public/Private Finance and Development: Methodology, Deal Structuring, Developer Solicitation*. New York: John Wiley & Sons.

Teisman, G.R. & Klijn, E.H. (2000). Public–private partnerships in the European Union: officially suspect, embraced in daily practice. In: S.P. Osborne (Ed) *Public–Private Partnerships: Theory and Practice in International Perspective* (pp. 165–186). London: Routledge.

Winch, G.M. (2002) *Managing Construction Projects*. Oxford: Blackwell Publishing.

Yescombe, E.R. (2002) *Principles of Project Finance*. London: Academic Press.

Zhang, X.-Q. (2004a) Improving concessionaire selection protocols in public/private partnered infrastructure projects. *Journal of Construction Engineering and Management,* **130**, 670–679.

Zhang, X.-Q. (2004b) Concessionaire selection: methods and criteria. *Journal of Construction Engineering and Management,* **130**, 235–244.

Further Reading

Akintoye, A., Beck, M. & Hardcastle, C. (Eds) (2003) *Public–Private Partnerships: Managing Risks and Opportunities.* Oxford: Blackwell Science.

Bennett, R.J. & Krebs, G. (1991) *Public–Private Partnership Initiation in Britain and Germany.* London: Belhaven Press.

Bult-Spiering, M., Blanken, A. & Dewulf, G. (2005) *Handboek PPS.* Utrecht: Lemma.

Dunn, J.A., Jr (1999) Transportation: policy-level partnerships and project-based partnerships, *American Behavioral Scientist,* **43** (1), 92–106.

Knibbe, A. (2002) *Publiek Private Samenwerking.* Alphen aan den Rijn: Kluwer.

Li Bing & Akintoye, A. (2003) An overview of public–private partnership. In: A. Akintoye, M. Beck & C. Hardcastle (Eds) *Public–Private Partnerships: Managing Risks and Opportunities* (pp. 1–24). Oxford: Blackwell Science.

Nijkamp, P., van der Burch, M. & Vindigni, G. (2002) A comparative institutional evaluation of public–private partnerships in Dutch urban land-use and revitalisation projects. *Urban Studies,* **39** (10), 1865–1880.

Webliography

European Union www.europa.eu.int (accessed June 2005)

Federal Acquisition Regulation www.arnet.gov/far/ (accessed July 2005)

HM Treasury www.hm-treasury.gov.uk/ (accessed January 2005)

National Council for Public–Private Partnerships www.ncppp.org (accessed January 2005)

4 Concessions

Nowadays, most western governments depend on private capital to develop large hard and soft infrastructure projects. The private finance initiative (PFI) form of procurement or concession contracting is one government approach by which private parties are increasingly involved in infrastructure projects. As mentioned in previous chapters, there is a great deal of confusion about the definition of public–private partnership (PPP). Private finance initiative, concessions and privatization are often used synonymously. In many countries, no distinction is made between the motives to start with PFI, PPP or privatization. In this book we distinguish between concession and joint-venture PPPs because the rationale behind and impacts of these different forms of public–private governance differ strongly. This chapter gives an overview of the motives, development and added value of concession arrangements, some illustrative examples of which are given in Chapter 5.

4.1 Concept of concession arrangements

Throughout the world, many types of PPP have been developed as governments have sought to increase private sector involvement in the delivery of public services. Concession arrangements have been of particular interest both in developing and developed countries, for the provision of different forms of infrastructure.

Here, we use the terms PFI and concession PPPs as synonyms. Concession arrangements incorporate many agreements such as design–build–finance–operate (DBFO), design–build–finance–maintain (DBFM) and build–operate–transfer (BOT): for all these variants the duration of the contract is long and maintenance and exploitation are outsourced. The concession arrangement differs from conventional projects in three different ways (Ball et al., 2000):

(1) The private sector organization involved not only constructs the capital asset but is also responsible for its operation and maintenance. In other words, the private client has a responsibility over the life cycle of the asset. A concession contract integrates design, construction and maintenance

(2) An output specification is used in which public sector clients define the services required

(3) Risks are transferred from the public client to the private sector. In concession contracts a 'genuine transfer of risk' to the private sector contractor

takes place to secure value for money in the use of public resources (Allen, 2001).

Life-cycle responsibility, output specification and risk transfer are the three typical characteristics of and motives for the concession arrangements, but can also be seen as the three major bottlenecks in the functioning of the partnerships. In practice, as we shall see here and in Chapter 5, the life-cycle value, the detailed requirements or output, and also the risks involved, are hard to specify and consequently to manage during the contract period.

It is important to distinguish concessions from privatization, contracting-out and joint ventures. Under a concession arrangement, the public client delegates to the private sector the right to provide a service. A public service is more or less contracted out to the private sector. It differs, however, from typical forms of contracting out in that the private sector provides the capital asset together with the services (Allen, 2001). It differs from joint ventures in the way that the public and private sectors are not 'real' partners during the project; rather the service and risks are 'transferred' from the public client to private contractor. In joint ventures, the government and private sector are both shareholders in a joint company; in a concession arrangement this is not the case.

The client retains control over the sector by incorporating appropriate terms and conditions in a concession contract or licence (Guislain & Kerf, 1995). Unlike in privatization, the public sector retains a substantial role in concession projects, with the private sector providing both capital assets and services (Romeiros de Lemos, 2002). The government retains responsibility for deciding which public sector services are to be provided as well as the quality and standards of these services (Akintoye et al., 2003).

Concession arrangements have been embraced by many countries and applied in different sectors. Concessions can be found in North and South America, Asia, Africa and Europe. In general, the main drivers in all these countries have been identified as budget deficits, ageing or poor infrastructure and growing demand for public services (Ahadzi & Bowles, 2004). However, as we shall discuss in this chapter, the motives and arrangements differ from country to country. The range of sectors in which concessions are applied has also been growing. Today, concession arrangements are applied in transport infrastructure, health care, education, prisons, culture and sport facilities, waste management, sewage treatment, social services and defence.

Beside the global attention concessions receive from governments, institutions such as the World Bank, United Nations, European Investment Bank and audit and planning offices in various countries are stimulating concession contracts in order to provide public services. The concept is also generating a fair amount of research interest (Akintoye et al., 2003; Ahadzi & Bowles, 2004).

In the next section we describe the historical development of concession arrangements in various countries and the current state of affairs. To understand the rationale behind concessions, knowledge of their history is important.

4.2 State of the art

In the past few decades many countries in the western world have applied concession schemes to cope with declining government budgets. In addition, the complexity of today's urban problems is increasing and public authorities are relying more and more on knowledge available in the private sector. Although the motives for launching concession projects seem similar at first sight, various approaches and procurement methods may be adopted. Despite the large number of projects, conferences and workshops on the added value of concessions, the discussion is often not a rational one, but a political and often emotional one. To understand the different motives and approaches, we have to look back at the introduction and development of concession schemes.

This chapter is primarily based on an international comparative study recently completed by the authors. The countries chosen to indicate the wide variety of approaches are the USA, the UK, Spain, Portugal and the Netherlands; developments therein are discussed in detail. Based on the analyses of these countries we are able to define conditions for success.

4.2.1 United Kingdom

Despite the relatively short history of PFI in the UK, many governments and private parties in the western world regard the UK as the front runner in concessions. Since the introduction of PFI schemes in the early 1990s, PFI practice has become well established. More importantly, UK service providers, legal consultants and investors are 'selling' the knowledge and experience gained in the UK to other countries. The UK model of concessions has been applied in many countries in the western world.

The development of PFI coincides with privatization; their introduction was therefore a cultural turning point. The stage for concessions was set in the 1980s when alternative forms of procurement were developed, such as outsourcing, privatization and concessions (Mustafa, 1999). The way in which projects were financed was being rationalized.

An important platform for the development of concessions was the Treasury's Ryrie rules of the 1980s which stated that privately financed projects could only proceed if they offered better value for money than a hypothetical public sector comparator, even if budget constraints meant that the public sector alternative would not go ahead (Hall, 1998). According to Allen (2001) the two fundamental principles of the rules were as follows:

- Private finance could only be introduced where it offered cost effectiveness
- Privately financed projects for public sector programmes had to be taken into account by the government in its public expenditure planning.

In later years, the Ryrie rules were relaxed so that the public sector comparator (PSC) would no longer be required if either the project could be financed through user charges or if there was no reasonable possibility of the project going ahead with the public sector (Hall, 1998).

The rules were superseded in 1992 by the PFI. In that year the Chancellor of the Exchequer (Minister of Finance) introduced new regulations that permitted the use of private capital for public sector products and services. For 300 years before 1992 the Treasury had opposed non-conventional ways of financing; then PFI was launched. Directives indicating the use of and need for the initiatives were given to the departments. In the various statements by succeeding Chancellors two major principles were stressed which characterize the rationale behind the PFI or concession contracts in the UK:

- The genuine transfer of risk to the private sector
- The generation of value for money in the use of public resources.

Because of the slow development of PFI schemes, in 1997 the Paymaster General decided to introduce new institutional and political rules to stimulate the use of concessions (Jacob & Kochendörfer, 2000). It was argued that the public sector could generate value for money when using PFI instead of traditional procurement methods. The motivation for the private sector was their ability to make a profit when using the long term contracts.

The year 1997 was an important turning point because the Paymaster General announced the end to universal testing – the rule that all capital projects had to be tested for private finance potential (Allen, 2001) and commissioned Malcolm Bates to carry out an independent review of the machinery of PFI. One of the review recommendations was to establish the Treasury Taskforce, which is responsible for 'road-testing' all significant projects and which reports directly to the Paymaster General.[1] The Treasury Taskforce was to be the central focal point for all private finance projects. The Taskforce focused on a number of significant projects, helping departments when negotiating with the private sector and with defining the output specifications to obtain value for money, and also published a number of guidance documents and statements on private finance arrangements.

The Bates report (1997) was followed by a second review in 1999. This report recommended the establishment of Partnerships UK; this was finally launched in 2000. According to Bates (1997) 'the ground breaking Partnerships UK will itself be formed as a partnership, with the private sector taking a majority stake in a joint venture with central government'. Partnerships UK consists mainly

---

[1] The Treasury Taskforce is divided into two sections: the Project Team and the Policy Team (Jacob & Kochendörfer, 2000). The Project Team consists of experts from the private sector and supports the project management of the departments by answering financial, juridical and other questions. The estimated lifespan of this team is two years. The Policy Team is responsible for contract conditions, execution and completion of significant PFI projects.

of experts from the private sector and only a small proportion of members from the public sector. The Partnerships UK organization is considered to function as a project manager who helps the public sector on a voluntary basis, providing expertise and advice for each PFI project and giving financial support (see www.partnershipsuk.org.uk).

On a central level, a local counterpart was established, the so-called 4Ps (Public–Private Partnership Programme) as a central institution to assist with procurement. The goals of the 4Ps are as follows:

■ To generate guidelines for best practice
■ To develop training programmes
■ To assist and evaluate the Gateway Reviews, which central government requires as a part of the application procedure for concessions
■ To influence central government policy with regard to procurement.

Engaging local authorities in a drive for improvement and high service standards was a central focal point of the UK Labour government policy in 1997 (Akintoye et al., 2003).

By early 2004, 219 PFI or concession schemes had been approved at a local level, of which 84 were operational at the time. Concessions have been applied locally in different sectors: waste, education, health and social services, leisure, transport and housing development. Approximately 90% of the private finance contracts are for schools and there are several reasons for such concessions being popular:

■ The contracts are for build and facilities management. There is little public opposition because teaching is not involved in the scheme. For many private parties, however, the fact that teaching is not included is the reason why it is difficult to generate value for money
■ The contract is part of the application to the Ministry of Education for the provision of schools. In education, there is a yearly application round on just one date; consequently, replication and standardization are possible, which makes bidding quicker and cheaper
■ Both central and local governments are stimulating the development of schools for which there is a strong demand.

In contrast to schools, there is no specific date on which to apply for funds in social services, which makes it more difficult to develop standardized contracts.

According to the UK government the establishment of the Treasury Taskforce, Partnerships UK and the 4Ps indicates the importance of knowledge transfer and professionalization of both the public and private sectors. The three institutions have developed a whole flow of standardized guidelines and public sector comparators. The strong focus on modernization and professionalization furthermore led to the creation of the Office of Government Commerce in 1999 which is responsible for modernization of the government procurement rules.

The National Audit Office (NAO) also plays an important part because it investigates the profitability of concession projects. The results of the investigations are published, as case studies, by the Stationery Office. According to the NAO, these results contribute to the standardization of the PFI-procurement process. Furthermore, the NAO monitors the tendering process and the contract procedures on a regular basis. Check-ups are made to see if the administration has made any mistakes. In combination with this investigation, the intended risk transfer should be determined.

The UK makes no use of unsolicited bids because of previous bad experiences. According to several key players to whom the authors have spoken, no real competition between bidders would exist if unsolicited proposals were to be implemented. In addition, it is argued that unsolicited proposals force private parties to become involved in politics.

Historical development in the UK shows that increasing value for money and the transfer of risks are the key motives for launching concession arrangements. Furthermore, the UK government sees professionalization of the market, standardization and knowledge dissemination as important conditions for success. Despite the level of experience with concessions in the UK, their impact on the way the public interest is safeguarded is often discussed in the media, most of the time not in a positive way. An important reason for the negative connotation is that concessions have been confused with privatization.

4.2.2 Spain

Concession arrangements have a long history in Spain, the first toll highway concession, the tunnel of Guadarrama, having been endorsed by the Act of 16 February 1953. The drivers for concession initiatives were quite different from those in other European countries. From 1950 to 1970 expenditure in the public sector was less than 25% of the gross national product (GNP). With this low rate of expenditure in the public domain, some public services could not be delivered adequately and other public services were left in the hands of the private sector (Torres & Pina, 2001). In those years, there was little control on cost efficiency and quality of the services provided. Nevertheless, even in later years when public sector expenditure grew to more than 40%, concession contracts were encouraged, with some short political interludes. In addition, the history of privately financed projects also shows how dependent the Spanish government is on private sector money. In the Plan Director de Infrastructures (PDI) of 1992 the total budget needed for the required infrastructure was predicted to be €171.3 billion for the period 1993–2007. This plan made it clear that 30% of the investments would have to come from the private sector. Unfortunately, private engagement was very small. Because of insufficient government funding, the PDI could be executed only partially.

An overview of laws since 1955 shows the development of concession policy in Spain. In 1960 Law 55/1960 set up standards for the construction, maintenance and operation of motorways. This law still offers favourable conditions to private concessionaires, such as government help in the form of subventions, deposits, adoption of exchange rate risks and tax advantages.

In 1965 the National Toll Highway Programme included the construction of 3160 km of toll highways. To stimulate the private sector, the decree of October 1965 included the possibility of providing a state guarantee to finance the new sections; concessions were given by decrees issued for each section and included advantages such as state guarantee, fiscal exemptions and advantages and the new model of 'exchange insurance' guaranteeing the remainder of the debt against exchange rate fluctuations.

In 1967 the Programma Autopistas Naçionales de Espana (PANE) was set up for the period 1968 to 1979, as a result of which 2000 km of toll roads were developed. The government supported the programme by, for example, taking over the risks of exchange rate fluctuations.

Law 8/1972 effected an important change in the policy on PFI. This law summarized and unified the advantages in each of the different laws and decrees that existed until then, and provided a clear and stable framework within which to operate. Law 8/1972 facilitated the use of various forms of private finance systems; because of this more highways have been procured in this way in Spain than in other countries. Government support was an enormous stimulus for the concessions boom in Spain. The law embraced previous arrangements such as fiscal profits, state guarantee, and 'exchange insurance'. The maximum period of concessions, however, was reduced from 99 to 75 years. Another important characteristic was that the activity of the concession was limited to construction, maintenance and operation of motorways. A big step forwards in the development of toll roads had thus been made (Jacob & Kochendörfer, 2000).

Motives for creating concessions as given in Law 8/1972:

- An increase in the legal security and economic efficiency of the concession system
- Greater flexibility in public works financing systems
- An improvement in the assignment of risks in public works concessions
- Positive effect on users: importance given to the phase of operating the works, quality indicators and the progress clause
- Greater private sector participation in the provision of public works
- Accumulation of experience in the field of public works concessions by Spanish concession-holding groups
- Enable the Ministry of Promotion's Infrastructure Plan 2000–2007 to be implemented, making infrastructure contribute to real convergence and full employment.

Law 8/1972 introduced 'financially free-standing projects' in which the consortium designs, builds and finances the infrastructure and manages the service. The operator then recovers the cost of the investment through the direct payments of the service user (Torres & Pina, 2001).

At the end of the 1990s Law 50/1999 was passed which stimulated the concept of shadow toll in infrastructure. In contrast to free-standing projects, the private sector recovers the investment through the sale of services to the public sector under the terms defined in the contract (Torres & Pina, 2001). The first shadow toll road in Spain has been endorsed by the Comunidad de Madrid. This is the M45, a 36.2 km ringway and is one of four shadow toll roads to be endorsed by the Comunidad de Madrid, the others being the M501, M600 and M50.

The political and economic climate in Spain was not always in favour of concessions. After the introduction of the Law 25/1988, for instance, several advantages for the concessionaires, such as adoption of exchange rates by the government and deposits, were dismissed. Consequently the development of infrastructure stagnated.

Research reported by Dewulf et al. (2004) stated that a study carried out by the Spanish building department in 1997 indicated the political motives for concessions. This study identified 1150 km of motorway that could potentially be operated by the private sector. Additional studies showed that with a concession period of 35 years, only 264 km of potential motorway would be profitable. The government therefore proposed another 442 km of new motorways that it could subsidize.

Law 13/1996 allowed concessionaires to enlarge their activities from construction, operation and maintenance of motorways to construction and operation of service stations and other real estate projects. To stimulate road construction, concession periods were lengthened from 50 to 75 years; taxes were lowered from 16% to 7%. Thanks to these policy decisions, private involvement in the planning, financing, construction and operation of infrastructure will continue and grow.

In contrast to many other European countries, little attention is paid to knowledge dissemination or centralized steering of the procurement system. Currently, no system of knowledge dissemination or general guidelines exists; nor does a national PFI unit. This results from decentralization; every region has built up its own knowledge independently. Although it may be politically difficult to create such a centre, it is highly advisable. Informally, there is a Political Economic Commission. At the Ministerio de Fomento (Ministry of Public Works) a special unit for the promotion of PFI exists; however, public sector representatives state that this is an engineering (planning and design) unit rather than a financial expertise centre.

4.2.3 Portugal

It is generally accepted that Portugal, of all European countries, has most enthusiastically embraced the PFI/PPP model for infrastructure finance, as pioneered in the UK. Privately financed projects are mainly carried through in the roads sector because of EU subsidies and credits with low interest rates from the European Investment Bank. However, these subsidies will probably fade out after 2006, when cuts in EU structural funds are to be expected.

In infrastructure, levying tolls is seen as a common activity. Projects in 2001 included the IP5 and Norte Litoral shadow toll roads, the Litoral Centro real toll road and a number of stadium-related projects for Euro 2004. The introduction of concessions is now being discussed in other sectors as well: the power sector may soon return to the fore, and it is said the government is keen to build PPP hospitals.

Many concession schemes have been developed in response to the Portuguese National Roads Plan (NRP) 2000. The NRP 2000 was launched by the former Junta Autonoma de Estradas (which in 1999 became the Instituto das Estradas de Portugal, i.e. the Portuguese Road Authority, within the Ministerio do Equipamento Social) and encompassed the road improvement and construction requirements in Portugal, defining the quality of service for each road forming part of the main or complementary network. It comprised a total of 16 roads, which are a mixture of shadow and real toll roads. Each concession included several sections that had to be built, enlarged, or merely operated and maintained. The roads were tendered under a DBF Operation scheme. In 1991 Portugal's road network included only 431 km of concessions; by 2006, it is planned to have a total of 2700 km of concessions in place, which will represent 90 per cent of Portugal's national highway network (US Department of Transportation, 2004).

In the first wave, the Portuguese government put out to tender two real toll roads and six shadow toll roads of which the first two were real toll roads; the contracts for these have been closed and concession timetables started. The six shadow toll roads (Sem Conbranca aos Utilizadores (SCUT) meaning without charge for users) were put out to tender in 1998. As of January 2002 all bar one of the contracts had been signed and/or closed.

Under the shadow toll system, the private operator does not charge road users, but receives direct payments from the government linked to the actual traffic level, thus enabling the government to fulfil its responsibility to provide roads while transferring the responsibilities for their operation and mainten-ance to the private sector, and hence improving efficiency. It was envisaged that by providing toll-free access to the motorways, and consequently avoiding any reluctance to pay by potential users, the system would transfer to the private sector a lower traffic risk, as the estimated level of traffic was relatively easier to determine.

Government payments are based on the number of vehicles using the road: the measurement unit is the vehicle kilometre and the tolls are structured into a banding system, whose thresholds in terms of vehicle kilometres and toll tariffs formed part of the bid submissions. An extra tariff (approximately double) is applied for heavy vehicles. In summary, the payments from the government depend upon three components:

- The number of vehicle kilometres counted on the road
- The type of vehicle (i.e. light or heavy vehicles)
- The level of shadow toll tariff, which is volume-dependent.

The level of shadow tolls to be applied (measured in euros paid per vehicle kilometre) depends on the traffic volume band into which each vehicle kilometre falls. A different toll is applied to each of the three bands. The only restriction from the government in the bid was that the tariffs could only decrease over time and that traffic flows above the top band would be assessed for each year starting from the beginning of the concession, and would be updated every year to reflect inflation. The concessionaire does not receive any shadow tolls for vehicle kilometres above a certain limit.

Since 2002, the government has been supporting the use of real toll roads in preference to shadow toll roads. Shadow toll roads are considered to deliver financial support from the government, which is, according to public agents in Portugal, not favoured in the EU (Dewulf et al., 2004). Moreover, public parties state that shadow toll roads are not financially attractive for the government, because in the past the government always had to pay more when traffic increased. The risks with the introduction of shadow toll are high and hard to predict at the beginning of the project. Therefore, Portugal has decided that well-used roads will be contracted out as real toll roads and others as shadow toll roads.

In Portugal, there is no central institute of knowledge, as in the UK with Partnerships UK. Because it was feared that knowledge about process approaches and process management would disappear when the administration changed, a central PPP government unit (as part of Parpublica, the unit of the Ministry of Finance responsible for privatization operations and outsourcing activities) was set up in September 2003. This unit is now working on the development of a public sector comparator. Procedures are becoming more streamlined with the establishment of this PPP unit. However, standardized procedures, governmental guidelines and centralized models and contracts are not yet available. To date, there has been no new legislation and concessions have been adapted to existing Portuguese laws.

In a concession strategy such as that developed by the Portuguese government, appropriate risk allocation is essential. Table 4.1 describes the distribution of responsibility associated with the risks of the Portuguese strategic plan (US Department of Transportation, 2002). The risk-control strategy suggests that the party best able to manage the risk bears it. For example, the risk of planning remains with the government and is the only risk it maintains in full. The risks associated with design, construction, operation and maintenance, latent defects and legislation are assigned to the concessionaire, while there is shared responsibility for environmental, land acquisition and force majeure events. There is a shared risk for revenue in the shadow toll method (US Department of Transportation, 2002).

Table 4.1   Risks and responsibility in Portugal (US Department of Transportation, 2002).

Risks	Responsibility		
	State	**Shared**	**Concessionaire**
Planning	X		
Design			X
Environmental		X	
Land acquisitions		X	
Construction			X
Operation and maintenance			X
Revenue (traffic)		X[a]	X[b]
Latent defects			X
Legislation			X
Force majeure		X	

[a] SCUT concessions (shadow toll).
[b] Real toll concessions.

One of the major challenges is how to cope with environmental issues. In the past, the Portuguese government used to procure before all environmental procedures had been completed. This resulted in many corrections afterwards and consequently in high costs for the government. Environmental problems may result in delays in design and construction, and a delayed commencement of tolling, in which case the government has to compensate the concessionaire for additional design costs, additional consultant costs and increased costs of environmental compliance, including land cost and cost of improvements. Consequently, the Portuguese road authority prefers to obtain environmental approval before launching any programme.

Right-of-way acquisition cannot be completely delegated to the concessionaire because only the government may exercise expropriation (condemnation) rights. In practice, the concessionaires can participate in the acquisition process, doing everything up to the determination of need. The first Portuguese concessions gave primary responsibility for acquisitions to the government, with parcels being identified by concessionaires and the government undertaking the acquisitions, but this method has proved burdensome. The most recent concessions have involved a transfer of significant right-of-way risks to the concessionaires, by transferring more and more of the expropriation activities to them. The concessionaires handle negotiations and the government provides the public interest declaration. If contested, the matter goes to court and the government handles the case. The potential for delay in the court proceedings is a government risk. The government also bears the risk associated with any requirements to acquire property outside the original corridor because of environmental demands. In some cases the government started acquisition proceedings early and ultimately discovered that the parcels concerned would not be needed for the final project configuration.

4.2.4 The Netherlands

An important impulse for the launch of concession contracts was formulated by the Kok administration in 1998. The second Liberal–Social Democratic administration led by Kok was confronted with insufficient public funds to meet the enormous investments needed to improve the transport infrastructure. Consequently, private contributions were considered as a possible solution and private finance projects were put on the political agenda (Koppenjan, 2005). A number of potential projects were mentioned, including several highways, the second Maasvlakte (expansion of the Rotterdam dockland area), high-speed railways from Amsterdam to Antwerp and to Germany, and the Betuwe route (a new railway for transportation of goods between Rotterdam and Germany).

As in the UK, the Dutch government stressed the importance of value for money; that is, private finance arrangements should be chosen if the bid of the private consortium offers better value for money than the public sector comparator. Comparable to the UK, the Dutch government has developed a tool for the evaluation of both the concession project and for the selection of the concessionaire. The public sector comparator (PSC) compares the net present value (NPV) of the concessionaire's proposal with the cost of design, construction, maintenance and operation using traditional methods. In this manner, the agencies can compare not only alternative concessionaires' proposals, but also the traditional procurement method (PPP Knowledge Centre, 2002; US Department of Transportation, 2002).

In summary, one could say that the rationale behind the Dutch concession policy is very similar to that of the UK. However, the development of the Dutch concession arrangements did not boom as in the UK. Moreover, Koppenjan (2005) concluded that despite the political and academic attention concession arrangements received in The Netherlands, concession contracts aimed at the development of infrastructure projects stagnated.

While concession arrangements have stagnated in the infrastructure sector, they have been launched for the construction and maintenance of public buildings, such as the renovation of the Ministry of Finance and a school project (see Chapter 5 for a description of this case). At present the Dutch government has identified four concession projects with an investment value of €300 million. These include a tax office, a courthouse, a prison and a deportation centre at Schiphol Airport, all projects of the Government Buildings Agency. Tendering is expected in 2006 and the government expects a cost reduction of 18%.

Another similarity with the UK is the focus on knowledge dissemination. Comparable to the Partnerships UK, a PPP Knowledge Centre was established to initiate and stimulate partnerships; it advises government agencies and provides private organizations with general information.

4.2.5 USA

Historically the private sector has played an important role in the construction, financing and operation of infrastructure. In the 1800s private companies were already building roads that were financed by tolls, but private involvement subsequently declined due to competition from railroads and greater state and federal involvement in the construction of tax-supported highways (General Accounting Office, 2004). In the twentieth century several plans for more private involvement were developed, but then abandoned. In 1956, for instance, the Federal Aid Highway Act established a tax-supported road system with revenues from motor fuel taxes rather than from tolls; tolling was prohibited on interstate highways (General Accounting Office, 2004).

Although the role of the private sector in highway financing and operation declined in the mid-part of the twentieth century, in the late 1980s, private sector involvement in these cases re-emerged. The role of the private sector increased because of (a) highway funding constraints at both federal and state level, and (b) a growing need for highly efficient surface transportation systems. Accordingly, legislation was adapted. The Surface Transportation and Uniform Relocation Association Act of 1987, for instance, allowed tolling on non-interstate roads. Activities for which public agencies were responsible have been transferred to the private sector. Currently, private sector involvement ranges from the maintenance and operations of individual highways or large highway networks to managing the financing and procurement of large highway capital expansion programmes (US Department of Transport, 2004).

Around 2000, US government budgets showed large surpluses together with predictions of excellent long term balances and declining debts: in this environment, the demand for concessions was low (US Department of Transport, 2004). At the same time, resistance to privately financed constructions was growing. Many people in the public sector saw little need to change their traditional ways of operating. Institutional inertia and misunderstanding of the private sector's methods of operation created fear of poor performance quality, job losses and a host of other supposed ills; resistance to concessions was manifest. According to the General Accounting Office (2004) there is significant political and cultural resistance to toll roads. Besides, traditionally, state and local governments build and finance highway projects through their capital improvement programmes. Of the costs of projects eligible for funding 80% were paid for by federal grants, which consequently provided a powerful incentive to build un-tolled roads (General Accounting Office, 2004).

The political commitment for concession contracts further declined after 11 September 2001, with the terrorist attacks only adding to the shifting balance. Rising costs for the war on terrorism, improved security measures and relief for the victims (both individual and corporate) further diminished the fiscal resources of the government. In addition, the firefighters and rescue personnel at Ground Zero and the Pentagon, and a dynamic president and mayor taking forceful leadership roles all created powerful public images. This effect was transmitted to many branches of government, leading the public to view the government in a more positive light. This included seeing the government as a more credible source of services to accommodate a whole host of public needs and concerns. All this has caused a shift in priorities and has forced many traditional sectors of interest to PPPs to be given a lower priority (Norment, 2002).

There are many US policies on private financing, with large interstate differences in legislation and policy on private finance being evident. Consequently, there is no federal knowledge centre for the dissemination of knowledge. Many states have legislation inhibiting concession contracts; for instance, some states prohibit DB or outsourcing certain agency functions. As of February 2004, 23 states have legal authority for private sector participation in transportation projects, of which 20 states have the legal authority to utilize participation of the private sector in highway projects (General Accounting Office, 2004). This could mean 100% private investment or sharing the costs with the state. The development of projects using tolls may require special legislation. In most states, for instance, the authority to develop toll roads is limited to special public authorities; new enabling legislation may be needed to expand and transfer these powers (US Department of Transport, 2004; www.fhwa.dot.gov/ppp).

Another difference between states is the legislation concerning unsolicited proposals. Some states solicit projects whilst others allow unsolicited proposals which encourage innovation and new ideas in both design/construction and financing. Among the 50 states, Virginia has the most advanced ideas in this area.

4.2.6 Other countries

In Germany, PFI schemes have been difficult to develop because of several legal barriers. Another obstacle for the success of concessions in Germany is the fact that the government is decentralized, with many differences in policies between Länder and municipalities. However, in recent years, legislation has been passed in order to support new public–private arrangements and several Bundesländer have installed task forces to promote concession projects. Several projects have been implemented, mainly school projects, one example of which is the Monheimer PPP Modell: Kirchner, a German developer, won the contest for the renovation and maintenance of eight schools in NordRhein Westfalen for a period of 25 years. However, the first and largest school PPP was in Kreis Offenbach and included the refurbishment of 90 schools and 100 buildings; according to Elbing and Jan (2005) this project could not be called successful for several reasons: the bids (two bundles were procured) were not well prepared, the bidders did not get accurate information for the bid, no standardized contracts were available, and the projects were not specified well enough. Consequently, objective bid evaluation was difficult. The first concession project in hard infrastructure in Germany was the tunnel crossing the river Warnow near the city of Rostock.

Political support for concessions has been absent for a long time in Italy. Clear political commitment was established in 2000 with the development of the Unita Tecnica Finanza di Progetto (UFP) as a central taskforce. Elbing and Jan (2005) mention several reasons for this change in political commitment: funding restrictions for infrastructure, problems in achieving the Maastricht convergence criteria and successes with concessions in other European countries.

In France there is a long tradition of private sector involvement in the construction and maintenance of infrastructure. As long ago as 1554 a contract was granted for the construction of a canal. Not surprisingly, the majority of the motorways in France are built and maintained as concession arrangements. The toll system means that almost all the construction costs are borne by the user. In 1998 tolls accounted for 65% of capital motorway costs. There are nine concessionaires, only one of which (Cofiroute) is fully private; central and regional government bodies comprise the remaining eight regional concessionaires through limited liability companies: Societés d'Economie Mixte (SEMs). The relationship between the public owner and the concessionaire is very well defined, as are the roles of the engineer and contractor. French legislation on concession stresses the importance of clear output specification and the needs of the public sector client which needs are defined in a programme. A French law of 1985 determined that the client must participate throughout the entire process. The government, furthermore, has carefully determined the appropriate risk allocation. In general, the risk strategy is very comparable to the Portuguese system (www.international.fhwa.dot.gov).

In Asia, the rise of private finance schemes was stimulated due to the growing needs of the increasing population and their even faster growth in expectations. This was further encouraged by dwindling governmental coffers, surplus private resources, and a search for efficiencies in providing infrastructure (Kumaraswany & Morris, 2002). In the past decade many governments in Asia have stimulated private investments in infrastructure development, telecommunications, and water management through different forms of BOO, BOT or concession arrangements. Under a BOO contract a private contractor will develop, finance, construct and operate a project on the basis of an off-take contract under which the government will agree to purchase the output of the facility (Tiong & Anderson, 2003). A BOT is quite common, with the difference that the contractor agrees to transfer the service upon expiration of the contract. In the mid-1990s many Asian countries launched new legislation to stimulate private investments (Li Bing & Akintoye, 2003). Hong Kong, in particular, has a substantial track record of concession contracts. The development started in the late 1960s with the first cross-harbour tunnel. In China, several state-approved BOT projects have been awarded since 1996. The Chinese government has, in the past decade, been developing the BOT infrastructure and refining concession protocols. Several pilot projects have been launched recently. In Vietnam, legislation on BOT was introduced in 1992; in Korea an act was passed in 1994 which enabled legislation for private infrastructure development (Li Bing & Akintoye, 2003). In Japan, the

New Comprehensive National Development Plan presented in 1969 advocated PPPs due to the lack of government funds. In the 1970s the number of concessions rose steadily. According to Bongenaar (2001) in 1993 a peak was reached of 400 concession projects in that year. The development of concessions in Japan was also stimulated heavily by the Fifth Comprehensive National Development Plan in 1998. Because the national debt as a percentage of GNP soared to more than 8% in 1999, financing of large infrastructure projects through BOT and BOO schemes was suggested (Bongenaar, 2001).

An overview of Asian development shows that many governments have been adapting legislation to stimulate private sector involvement in the provision of public services. Moreover, as is the case in the UK and The Netherlands, some Asian governments installed a central knowledge centre on concessions. For instance, in the Philippines, a BOT centre was installed for the planning and implementation of private finance related policies. This centre provides information and technical support (Chang & Imura, 2003).

Most of the large-scale projects in Asia are financed using so-called non or limited recourse project financing (Tiong & Anderson, 2003). This is a form of debt financing in which lenders rely exclusively on the revenues generated by the infrastructure project as a source of loan repayment. Due to the strong economic growth there is a large need for infrastructure investments. But governments in China, Taiwan, the Philippines and other Asian countries are facing large demands for spending in sectors such as health care and education that lay a burden on the infrastructure investments. Because of the pressure on public financing, private sector involvement in infrastructure will continue in the future.

Australia also has a long history in concessions. In the early 1990s many concession projects focused on toll roads, hospitals, water and power. In the mid-1990s concession projects were introduced for prisons, sea ports and sports stadiums, and in the late 1990s also for airports. Early twentieth century concession contracts were signed for defence projects, schools and courthouses (Crump & Slee, 2005). The state with most experience with concessions is Victoria. Comparable to the UK, the method of evaluation is heavily discussed, because it is biased in favour of concessions.

4.3 Value for money and motives

Governments worldwide have implemented concession contracts to cope with increasing demands for public services and decreasing public budgets. The level of political and public acceptance differs from country to country. In Spain, there is a general belief that concessions, especially in infrastructure, are good for society, but applications for hospitals and schools are discussed in the media. In the USA, on the contrary, concessions are a politically tense issue. For example, the Public Citizen Community of Interest (Ralph Nader) states that concessions are in fact private actors who want to take over and control public matters, and who want to take over public personnel and fire them afterwards. These are strong, emotionally charged issues that work against concessions in the USA.

In general, one could say that concession contracts for highways and railways are politically and publicly accepted because in those cases there is often an alternative road available. Toll bridges are acceptable because the opposed quality and service level complies with the demands. Introducing tolls on existing roads or bridges, however, is less publicly accepted. Concessions in health care, education and other social infrastructure are less widely accepted.

In political statements all over the world abstract goals are formulated to substantiate the concession policy, such as:

- Investment in infrastructure: long term and sustainable improvements
- Making better use of public money
- Better value for money
- Focus on the life cycle.

Achieving the concept of *best value for money* is the overall political driver to launch concession arrangements. Best value is, however, a relative notion that refers to the 'optimum outcome of a business process' (Akintoye et al., 2003). When asking project managers how they defined value for money the following answers were given (Dewulf et al., 2004)

- Concessions lead to the dismantlement of governmental debts and jobs
- Projects could make a quick start whereas with the traditional approach they would have waited
- PFI generates additional sources of income in the private sector
- Clear restriction and transfer of risks onto the private sector
- Increase of efficiency because planner, builder and operator are co-operating (all the important partners in the life cycle have to bring in risk capital)
- Innovative ideas and the management knowledge of the private sector are used optimally.

What constitutes the best value depends on the motives and interests of the governments and may change over time due to political, economical and social developments. It is then also difficult to assess the performance and consequently to compare the performance of concessions with those traditionally procured. Historical analysis shows that various motives led to the boom in concession arrangements; some motives were similar whereas others differed strongly.

In general, the emergence of concession concepts in most countries could be seen as a reaction to government needs for funding and as a new way of governance. Public sector financing of large-scale projects has become difficult in the past few decades and governments are increasingly relying on private sources. Besides financial and economic reasons, many scholars refer to more political and social motives in launching PFI schemes, referring to the importance of a greater involvement of private parties in planning and development in order to avoid implementation problems, incorporate managerial and technical knowledge of private parties and generate long term commitment to public plans. Involvement of key stakeholders, for instance, creates the potential for planners to expand their understanding of urban problems and to develop a stronger set of policies for dealing with urban issues (Innes, 1998; Beierle, 2002; Graaf, 2005). In addition, involvement could create a sense of ownership over a plan's content and reduce potential conflict during its realization (Wondolleck & Yaffee, 2000). Forester (1989) and Graaf (2005) have also stressed that organizations bring valuable knowledge and innovative ideas that could increase the quality of the adopted plans. One example of valuable knowledge is management expertise. Private actors can often act more quickly and more efficiently than public actors. They have larger financial strength and knowledge of relevant markets.

The concept of value for money also refers to the whole life value of the service provided. Traditional contracts do not stimulate contractors to consider whole life costs in their bids. In concession projects, however, contractors are more aware of the whole life costs because they have to calculate the costs of operation over a concession period in their bid (Flanagan & Jewell, 2005). Since owners and users are increasingly concerned about the costs of maintenance and operation, whole life appraisals are becoming more common.

In comparison with traditional procurement schemes, transaction costs for both public and private partners are very high. The question is whether these costs counterbalance the financial benefits of this extra competition. However, the up-front investment could pay off when it comes to agreement on policy and implementation. According to Brody et al. (2003), involvement is needed to produce enduring plans. Avoiding potential disputes is becoming more and more important because many large projects have resulted in big claims on construction contracts. Private finance approaches minimize claims because they completely restructure the risks involved; there are, according to Scott & Harris (2004), fewer opportunities for claims to be raised.

The US Department of Transportation (2002) gives a good summary of the financial, political and economic advantages of PFI in Table 4.2. Despite the more intangible motives mentioned in the literature, economic and financial motives dominate the political debate. Moreover, in practice, decisions to start a concession scheme are often based on purely financial considerations. In most countries, governments are using instruments such as the public sector comparator to judge the value of an integrated contract: this comparator is often purely financial.

One could surmise that the general political motives for launching concession schemes are quite similar all over the world; however, the accents or immediate causes are quite different and consequently the concession approaches, as well as the organizational, legal and financial frameworks, differ strongly. As the historical analysis showed, one major reason for the different strategic choices is the difference in cultural and historical background between the countries. The Dutch concept of PPP, for instance, originated in the 1980s in inner-city projects by reason of social surplus value (Smit & Dewulf, 2002). In the UK, the

Table 4.2 Financial, political and economic advantages of PFI (US Department of Transportation, 2002).

Financial advantages	Economic and social advantages	Political advantages
Easing of budgetary constraints	Streamlined construction schedule and reliable project implementation	A new role for the public authority
Optimal allocation and transfer of risk to the private sector	Modernization of the economy and improvement of services	Allocation and not 'abdication'
Realistic evaluation and control of costs	Access to financial markets, combined with the development of local financial markets	Project stability

PFI was launched to combat budget-crosses and delays in schedules, and to mobilize private wealth. The emergence of concessions in Spain can be explained by its historical tradition of toll roads and its genuinely accepted norm that the user pays.

The mix of cultural, historical, economic and political developments determines the concept of private finance arrangements or concessions. However, the choice for a certain approach is not always a rational one, but may be determined by coincidences. The fact that the concession approach in Portugal is very similar to the British one is primarily because an English financier and UK consultants were involved in the first Portuguese concession project. Table 4.3 summarizes the motives for concessions in Spain, Portugal, the USA, the UK and The Netherlands.

When talking to public and private parties involved in concessions we find that a wide variety of added values are expected. Our international study revealed that there were differences not only between countries but also between the expectations of the public agents and private parties involved, as well as between the various sectors. For the countries analysed we will now elaborate on the various motives.

4.3.1 United Kingdom

The political motives for starting concessions are twofold: to increase the number of projects, and for financial reasons. According to the Treasury, concessions are only used if they can meet government requirements and deliver clear value for money without sacrificing the staff's terms and conditions of employment. In assessing in which circumstances concessions may be appropriate, the government's approach is based on its commitment to efficiency, equity and accountability, and on the Prime Minister's principles of public sector reform (HM Treasury, 2003). According to the Association of Chartered Certified Accountants

Table 4.3 Overview of motives to introduce concessions in the provision of public goods (Dewulf et al., 2004).

Specific motive	Spain	Portugal	USA	UK	The Netherlands
Stimulate economic growth	++	++	–	–	–
Urgent need for infrastructure	++	++	–	–	–
Greater efficiency in use of resources	–	–	–	+	+
Improved service quality	–	–	–	+	+
Generating commercial value from public assets	–	–	–	–	–
Financial need	+	++	–	+	+
Need for innovation	–	–	–	+	+
Time and cost savings	–	++	+	++	+

Key: ++, very important; +, important; –, not important.

(2004), PFI is not an alternative to the public sector capital programme, but in practice PFI is too often considered the only option.

Motives may differ slightly from sector to sector. In our study the following motives were mentioned by various players:

- *Department of Health* For major new facilities, a concession arrangement is the primary choice. In 2002, the Prime Minister said that PFI policy should play a central part in modernizing the infrastructure of the National Health Service (NHS)
- *Treasury* If a project is considered necessary and no public money is available, the government decides to choose a concession
- *HM Prison Service* Concession contracts are used when new prisons need to be built. It is also possible to use them for the maintenance and renovation of prisons. However, concessions cannot be used when adding to or expanding existing properties because legislation does not allow this. If possible, HM Prison Service will not favour traditional contracts, but rather design–construct–maintain contracts. (External) finance as in DBFM contracts is not always necessary or possible.

Added value differs between the public agent and private sector. UK public agents argue that concessions generate:

- Better value for money
- Fewer risks
- A life cycle approach
- Greater speed of delivery.

In the interviews we held, private parties, on the other hand, defined added value as:

- Discipline in delivering on time and budget
- Innovative ways to win work

- Contractors becoming more commercially aware
- Movement from high risks, low margins and short term, to spread of risks, better margins (because of the involvement of facility management) and long term.

4.3.2 Spain

In the 1960s and even today the motive to start concession arrangements was and remains primarily financial. In the 1980s, during the administration of the socialist government, there was a pause in this policy with significant investments in infrastructure which resulted in a huge deficit. Today, the government is again promoting mixed financing. Research results show that public and private partners are very happy with this system.

The following motives underlying the concession policy can be distinguished:

- Necessity
- Operational effectiveness due to mixed financing. The administration is based on French law, which involves many regulations and procedures; mixed financing is much easier to operate
- More transparency and more flexibility.

In our interviews, the Ministry of Transport stressed that procurement through concessions was not always a better governance structure than traditional ways of dealing with the market, but the concessions contract was often chosen for two reasons:

- Financial constraints of the government
- The user pays, in which case an alternative road has to be available.

According to the Ministry this policy is largely accepted in public opinion.

When there is no alternative road, a shadow toll highway will be developed instead of a regular toll highway. Other reasons that are mentioned for the introduction of shadow toll roads are:

- Avoiding traffic jams and stimulating use of the roads for which the user does not have to pay
- Reducing risks
- Stimulating the private sector to participate in the development of road projects.

According to public parties interviewed, the added value of concessions is:

- Cost minimization, especially savings in road maintenance
- Greater efficiency in the process, but it is questioned if the costs were lower and was acknowledged that this is hard to evaluate afterwards
- More transparency and more flexibility. The government stated that flexibility in the contract is required, because if the public interest changes, the government has to be able to break the contract

- Using concessions certainly creates more discipline among public and private parties when working together. Nowadays, it is generally recognized that government resources are not limitless.

An important by-product indicated by the Ministry of Transport was that users have to pay tax on tolls, which means extra revenues for the government.

Because of the experience Spain has gained with roads, evaluations or benchmarks may easily be made. The differences between highways and railways concern the time needed to realize them. For the railway sector, the partners involved are especially cautious because there are many risks that need to be controlled.

According to private parties, the Spanish system has several advantages. First, there is a restricted bidding process time frame: no pre-qualification, four months for tender, four months maximum for award. According to Cintra, their estimated costs for bidding are €0.5 million. In the UK, Ireland and Portugal a bid takes two years because the government asks for confirmation of all financial partners, and external lawyers are needed. In Spain when a company wins the bid, in general it has one month to pay the government (in this sense the financial close has to be made after the bidding process). Second, the Spanish concession approach contains clear awarding criteria: points are allocated on the basis of specific technical (maintenance, level of service, electronic display panels) and commercial criteria (including toll level, concession period, grant). The tariff is set to a maximum level (average circa €0.07/km). There are also some conditions, such as the toll ports should be able to manage traffic jams. Another condition consists of criteria for the quality of the road. A third advantage is that the requirements of the authorities are as follows:

- No need for financial closure when the contract is signed
- No authority direct agreements; no pre-contract signature approval of design and construct contract; no need for operation and maintenance contract.

Additionally, the technical and financial capabilities of the consortium are included as part of the submission. There is no concession contract negotiation process. The concession contract is known by all bidders and remains unchanged from process to process. This approach is convenient for bidders. Furthermore, the authorities are responsible for the planning and pre-design, including environmental approval, and public information process. A detailed plan is made by the authorities. Finally, it is a very transparent system; every bidder can have a look at the offer of the other bidders.

4.3.3 Portugal

The primary motives for the Portuguese concession plan were: first, Portugal's entry into the EU; second, the need to strengthen its trading ability; third, to control the budget because Portugal exceeded the 3% deficit rule of the EU. Concession arrangements gave Portugal an opportunity to reduce its budgetary deficits and consequently to attain the criteria to enter the monetary union. Hence, it could be argued that the main incentive for concessions in Portugal is purely financial.

Other reasons found by our research are:

- The importance of true competition, it being argued that procurement of concession schemes stimulates competition in the market
- Lack of knowledge on the public side
- Speed of delivery of services.

Recently the motives have changed. The Parpublica PPP unit was established partly because the government became aware that, due to the concession schemes, payments were being postponed. Today, concession proposals are evaluated on the subject of efficiency.

One important change in recent years has been that whereas in the past the sector minister had been primarily responsible for the concession arrangements, recently the Minister of Finance has become more involved, as indicated by the establishment of the PPP unit as part of Parpublica of the Ministry of Finance.

The disadvantages for the government reside in the fact that the negotiating processes with the private sector are difficult. In addition, for some years there has been a trend for the skilled persons involved to leave the government and take their expertise with them.

For the private sector, added value could be situated in:

- Great freedom of design
- No pre-qualification
- Transparency in the selection process
- Technical innovation being possible.

The current Portuguese administration is in favour of introducing concessions in health care provision. Respondents were not convinced about the added value of introducing concessions in hospitals. Most respondents felt that the costs would not be lower in comparison with the traditional delivery of services. The motive to introduce concessions in health care is not primarily a financial one. In Portugal there is a doctor shortage and 95% of physicians are not on the payroll of a hospital. Introducing concessions is seen by the current administration as a way of stimulating efficiency and developing more efficient management.

4.3.4 The Netherlands

Despite the differences in motives between projects, we can argue that in the Dutch context budget constraints are not the principal motive for starting with concessions, but generating more value for money is. This can be translated more specifically into (see e.g. PPP Knowledge Centre, 2002, 2003; Regieraad Bouw, 2004):

- Innovation in physical concept and approach
- Greater efficiency
- Better value for the customer
- Creating a life cycle value.

Although the formal political arguments focus on the value added, a survey among public project managers of concession contracts in infrastructure revealed that the prime reason to start with the concession project was financial value (Bult-Spiering & Dewulf, 2002). In contrast, the reason for launching innovative public–private arrangements in urban projects is quality improvement. For these urban projects joint ventures, not concessions, were formed. Concession contracts are not preferred for urban inner-city projects because the public interest is quite diverse and cannot be safeguarded by means of standardized contracts, as is possible for infrastructure projects (Smit & Dewulf, 2002).

4.3.5 USA

According to Austin and McCaffrey (2002) the most common motive cited in the literature for the genesis of modern concessions in American cities is economic development, particularly in cities with a deteriorating inner core.

In our study, the dominant motives for concessions proved to be as follows:

- Obsolete properties
- Budgetary deficit
- Maintenance efficiency
- Under-utilized government assets
- Significant financial problems
- Great need for new investments.

In our research, respondents mentioned two challenges:

- Public PPP authorities must be extended
- Performance instead of lowest price should be the predominant selection criterion.

Both public and private respondents mentioned the following general advantages of concessions:

- Time saving
- Selection of the private partner on performance and value for money instead of selection on price
- Partnerships could improve the quality of the solutions.

Besides these advantages, some specific benefits were identified by the public respondents:

- Budget deficits are reduced and access to private sector money is gained
- Responsibility and reliability lies with one single source (i.e. the consortium)
- Risks may be shared with private companies.

Specific private sector advantages are:

- Pre-selection/pre-qualification means less competition and that the competitors are known
- More control over the total process.

4.4 Tendering and selection procedures

Chapter 3 described various procurement systems. In this section we will elaborate on the tendering procedure and criteria that are specific for concessions. The historical analysis described in the previous section revealed differences in motives and approaches. Consequently, differences in tendering and selection procedures can be distinguished. Two extremes can be identified:

- The Spanish system characterized by a short and transparent tendering procedure
- The Anglo-Saxon system characterized by a procedure that lasts almost two years. (Most European countries and many Asian countries are adapting the UK system.)

In this chapter we will describe the Spanish and the UK systems as snapshots of the two extremes. The Portuguese system will also be described because this country has adopted the UK system but with some important adjustments.

4.4.1 Spanish system

The Spanish system can be characterized as being quick, efficient and transparent for all parties. In Spain the bidding process has a restricted time frame. There is no pre-qualification, preparation for tendering takes four months, and the award may take up to four months maximum. The costs for bidding are estimated to be approximately €0.5 million. The technical and financial capabilities of the consortium are included in the proposal.

The awarding criteria employed are clear. Points are allocated on the basis of specific technical (maintenance, level of service, electronic display panels) and commercial criteria (including toll level, concession period and subsidies). Additionally, some conditions are given; for instance, the toll ports should be able to manage traffic jams. Another condition consists of criteria regarding the quality of the road.

The Spanish government has only limited requirements when signing off the contract. Thus, when the proposal is awarded and the contract signed, the concessionaire has one month during which to pay the government. Financial closure is made after the bidding process. In contrast, the bidding process in the UK, Ireland and Portugal takes up to two years. The government in these countries asks for confirmation of all the financial partners; external lawyers are needed to support this financial closure. There is no concession contract negotiation process: the concession contract is known by all bidders and remains unchanged from process to process, which is seen as a convenience for the bidders.

Before releasing a new tender, the following tasks are undertaken by the Spanish government:

- Preliminary design preparation
- Environmental studies and approvals required
- Public information period (in accordance with the Spanish legislation period)
- Financial feasibility study and European Investment Bank (EIB) consultation.

The government stresses the importance of transparency during the tender procedure. For instance, bidders are given the opportunity to review each other's offers for several reasons:

- Increase of market competitiveness. Bidders can learn from what others have developed and included in their offers, and use it in subsequent processes
- Complete review of the offers. Any inconsistency in the offer can be discovered by the competitors and notified to the government
- Complete transparency makes it possible to check clearly whether one offer is better than another, and if the points have been awarded accordingly.

The criteria used for the selection of the bids are as follows:

- Costs
- Design
- Maintenance
- Technical performance.

The importance of design and maintenance criteria will depend on the region or unit and on what core value/public value is wanted. According to public respondents, a review of the proposals of the bidders indicates a high level of technical innovation. The private parties, however, indicated that there is hardly any difference in technical performance between the bids of competitors; the major difference between the bids is their price.

Some examples of criteria used in tenders:

Toledo–Consegra Highway (shadow toll, regional government)
- Economic offer   400 pts
- Design and build   300 pts
- Operate and maintain   150 pts
- Financial structures   150 pts
Total   1000 pts

Airport link Madrid (central government)
- Technical quality   400 pts
- Financial standing and feasibility   250 pts
- Efficiency of concession design   150 pts
Total   800 pts

Private partners are involved in the project as early as possible. Early in the process the market is consulted and has maximum influence on the design. During the formal procurement procedure there is no communication between the government and private bidders to avoid collusion.

The concessionaire is entitled to ask for measures intended to re-establish a financially balanced budget for the concession (e.g. by requesting an extension of the concession period, an increase in the tolls and fees received by the concession holder, etc.). The new law sets forth a limited number of cases in which the contracting authority must take measures to re-establish such a balance, in favour of either the concessionaire or the contracting authority itself.

4.4.2 Anglo-Saxon model

The tendering process in most western countries takes longer than the Spanish system. As became clear from the historical overview, many countries have adapted the UK system.

United Kingdom

The tendering process in the UK consists of the following phases (Jacob & Kochendörfer, 2000):

(1) Determination of the needs: quality and the quantity of the desired services and the expected standards of effectiveness and efficiency
(2) Strategy decision: concession-capability of the single project, financial feasibility, social benefits, proposed support
(3) Outline business case (OBC): background and the expected results, public sector comparator (PSC), approval of responsible administration
(4) Advertisement in *Official Journal of the European Community* (OJEC)
(5) Pre-qualification based on project finance, experience, references and companies' due diligence: selection of a maximum of four candidates
(6) Invitation to tender (ITT): detailed offer, structured concession agreement
(7) Assessment of the received offers based on general criteria and tendering conditions
(8) Invitation to negotiate (ITN)
(9) Best and final offer (BAFO)
(10) Announcement of the winning bidders and notification of the 'losers'
(11) Contract closure.

The tendering procedures differ slightly from sector to sector. For instance, the NHS used to follow the procedure described in Fig. 4.1, but more recently has made an effort to speed up tendering. Especially in health care, the procurement time proved to be very high: from 22 to even 60 months in the period 1994 to 1998. However, in roads and light rail projects the procurement time was less than 20 months (HM Treasury, 2003). In the new setting, the NHS goes from four or

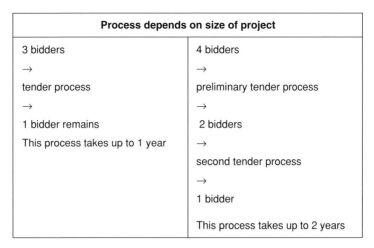

Figure 4.1   Tendering process for the National Health Service, UK.

more bidders directly to a preferred bidder; by so doing the bidding costs will be decreased and time saved.

An important characteristic of the Anglo-Saxon model is the use of a public sector comparator (PSC). The PSC is used to compare traditional procurement with the procurement by concessions and to define the price/quality ratio. If the bids are lower than the expected costs in the PSC, the procedure has to be repeated. The criteria of this important tool are:

■ Value for money
■ Affordability: comparing costs of public realization to the private offer.

Using the PSC forces the administration to calculate the total life cycle costs. Thinking about the real cost-effectiveness of a project has been big step forward for the public sector.

The PSC is filled in and checked during the course of the process. The PSC is set up in the initiation phase and should be ready before the invitation to tender. The outcomes of the PSC are not made available to the private sector, to avoid stimulating them to use this outcome as a target price. After selection of a preferred bidder, the outcome of the PSC is provided to them.

To explain the role of the PSC, an outline is described of the application procedure that local governments have to undergo when applying for central level approval of a concession tender. The following steps have to be taken by local government:

(1) Develop an outline business case that has to be approved by central government (as they provide the necessary funds) before procurement. This case is discussed by the central government committee in which the 4Ps takes part.

(2)   The outline business case must show:
  — Options appraisal
  — Value for money: capital value is estimated in the PSC. According to several public servants involved in concession arrangements, the offer of the private bid should be about 70% of the cost if a public agency were to provide the service (capital costs and operating costs). Value for money, however, is apparent not only in the comparison of prices, but also in comparison of the (division of) risks (Dewulf et al., 2004)
  — Affordability. Local government pays for the private finance contract, not as a lump sum but by spreading payment over several years
(3)   The final business case has to be approved by central government before the contract is signed.

The role and usefulness of the PSC, however, is hotly debated. According to Jeremy Colman of the National Audit Office who has publicly criticized the rigid use of the PSC, it is an opaque instrument. At a conference of the National Audit Office in 2003, he called the PSC 'pseudo-scientific mumbo-jumbo' and the figure produced by the comparator 'not information, just noise'. In his view, the discussion should focus on the risks involved and the benefits to be gained. To justify the concession option, departments have, according to the National Audit Office, relied too heavily on the PSC, which has often been used incorrectly as a pass or fail test. The outcomes of the PSC have been applied with a spurious precision despite the many uncertainties involved in the calculation and the fact that the numbers could have been manipulated to obtain the desired results. Before the concession route is chosen departments need to examine all realistic alternatives and make a proper value for money assessment of the available choices (Committee of Public Accounts, 2003).

The selection criteria used are often of a financial nature. According to the Department of Health, the selection criteria for the bids are based on EU criteria as deliverables offered, and financial robustness. Every set of criteria is looked into separately and brought together in a matrix. The most important deliverables are design quality and the quality of service delivery. Only if these deliverables are appropriate is the bid checked on price.

In general, in the UK the costs for a concession procedure are very high. Local government has to pay for procurement costs, which may amount to £0.5 million.

The Netherlands

Another example of the Anglo-Saxon procedure is found in The Netherlands. The Dutch system is very much like the UK one. The Dutch government has developed a PSC and incorporated it into the tendering procedure. The DBFM contract for the N31 road project provides an example of a selection process utilizing the PSC. The project involved improvements required for traffic safety and traffic flow reasons. The plan was to widen the road from one to two lanes. The road is approximately 25 km in length, and includes one bridge and an aqueduct. The estimated construction cost was €100 million (US Department of Transport, 2004). The Ministry of Transport determined the advantages and disadvantages of concession in a systematic and transparent way, using the PSC. The PSC was based on an estimate of the net present value, and the costs and benefits associated with the project. The study concluded that DBFM would be an efficient means of proceeding with the project, and also that there was no significant difference between DB and DBFM. Based on the PSC, the Ministry decided to proceed with the project as a DBFM pilot project. The proposed contract included 15 years of maintenance. In general, the tendering process follows the outline of the UK system and can be seen as a combination of requirements of Dutch law and the EU procurement directive; it includes the following steps (US Department of Transport, 2004):

- Advertisement: issuing an invitation to tender
- Pre-selection: three to five criteria for selection are determined based on issues specific to the project, including experience with DBFM, experience with design and construction of comparable projects
- Consultation (industry review): the Public Works Department provides draft terms and conditions to contractors, holds an initial meeting with all pre-qualified contractors, holds one-to-one meetings with individual firms, receives written questions from contractors, and provides written answers to all contractors
- Request for bids: including final form of contract, allowing alternative bids associated with risk
- Bidding phase: proposal includes design, risk analysis, draft quality plan and financial proposal
- Negotiations: selection of two bidders for negotiations
- Best and final offers: selection based on 'most economic bid' (best value) includes evaluation of design quality and the financial plan, including the net present value (NPV) of payments to be made by the Public Works Department.

The contract for the N31 includes incentives/disincentives based on road availability to encourage safety and minimize congestion. The contractor is subject to penalty points for not meeting safety regulations. If too many penalty points are received, payment will be reduced. If the contractor's performance still fails to improve, the government may even terminate the contract (US Department of Transport, 2004).

An important element in the Dutch tender procedure for infrastructure works is the systematic use of performance specifications. In traditional procurement the Ministry of Transport bases its contract on specifications of material behaviour and requirements for basic materials and processing. In concessions, however, the contract is based on more general criteria. The performance is then based on 'road-user wishes' or 'performance requirements', e.g. requirements for road surfaces, such as noise reduction and smoothness (www.fhwa.dot.gov/ppp).

Portugal

As mentioned in the previous section, Portugal has a very short history of concession arrangements, but is making up arrears. By doing so, Portugal has copied the UK system with some slight but important adjustments. Because Portugal is operating under the new EU rules and is involved in a significant number of concession agreements, the government has created a rigorous and repeatable selection procedure. The framework may be seen as a standardized procedure, but to date it has not been institutionalized.

The tendering process consists of two stages, with no pre-qualification in the first phase. In the first phase, the bidders are checked for their financial and economic stability as well as their experience. After the evaluation, two bidders are selected. In the second phase, the two bidders are invited to join the negotiating procedure. The procurement process gives the concessionaires five months to prepare their proposals following receipt of the request for tenders. The proposals are next presented to the sector department (in infrastructure, the Instituto das Estradas de Portugal) and, since 2003, also to the PPP unit of Parpublica. As the proposals involve design, construction, operation, maintenance and other services, the evaluation is quite complex and takes up to six months to complete. The evaluation is presented in a report on which the bidders get the opportunity to make comments. These comments are then summarized in a new report. The reports of the bidders are accessible to other parties.

The bids are evaluated by the following criteria:

- Costs (approximately 40% of the deal)
- Risk allocation: the risk is calculated by estimating the difference between the deals in an optimistic versus a pessimistic scenario
- Period of the concession
- Robustness of the proposal (legal and financial)
- Technical performance on construction
- Technical performance on operation.

Compared with the UK system, the major difference occurs during the negotiation period, during which the bidders bring their bank representatives with them. The technical aspects, as well as the financial and legal issues, are discussed at the same time. The reason for the involvement of the banks at this stage is to generate their commitment. In the UK, negotiations with the banks start after the final negotiations with the bidder/constructor. In contrast, in Portugal banks have to sign the contract/deal. In the past, some banks refused this early involvement, e.g. the Bank of America, but subsequently they agreed to this process. The Portuguese system stimulates banks to assess the risks and give their commitment

Table 4.4 Portuguese concessionaire selection process (US Department of Transportation, 2002).

		Duration (months)	Total period (months)
	Publicity		0
Bid preparation		5	
	Bid presentation		5
Evaluation		6	
	Two bids short-listed		11
Negotiation		4	
	Contract award		15
Finalizing legal requirements and contract terms		3	
	Financial closure		18

at an early stage of the procedure. It is a triangular process between banks, bidding company and government. According to the public parties, the tendering system in Portugal was very expensive at the start but is now becoming cheaper (Dewulf et al., 2004).

As mentioned above, the most important criterion in the procedure is the financial structure of the proposal. The evaluators believe that adjustments to the technical demands could be made during the negotiation with the bidders. In general, it is argued that it is difficult to use criteria other than price because of their being less transparent/objective.

In the first shadow toll projects, only two consortia submitted proposals. Today, six or seven consortia are bidding regularly in procurement procedures. The Portuguese tendering process, as shown in Table 4.4, is as time- and cost-consuming as it is in the Netherlands and the UK. The completion of the contract terms and contract award is accomplished in three months; the entire selection process takes an average of 18 months to complete. Today, the government is putting an effort into shortening the procurement system and hence into lowering the bidding costs and making it more attractive to bid.

4.5 Performance of concession PPPs

Most studies on concessions focus on a single element or performance criterion, on a specific country or sector, or on single case studies. As shown in Chapter 2, many interests are at stake in construction projects and an integrated assessment should be made over a long time period to judge the performance of concessions. Evaluating the performance of concessions is, however, a difficult task for several reasons:

(1)   Most criteria for best value are opaque or subjective and therefore hard to measure. For instance, the (societal) value of a project is important, but

judgement on the basis of societal value fails because its quantification appears to be difficult.

(2)   Many of the criteria can only be assessed over a long time period. Since the bulk of the concession projects were started in the late 1990s it is too early to make a final statement on the performance of these contracts.

In addition, assessing performance by analysing the conditions for success is even more difficult because a large variety of variables influence performance. This section is therefore not meant to be conclusive, rather explorative.

In Chapter 2, a distinction was made between process and product performance. We will further elaborate on these two performances. Product performance was defined in Chapter 2 as financial performance (revenues, cost-efficiency, transaction costs and risks) and content performance, whereas process performance refers to actors' fit, public interest and behaviour.

4.5.1 Product performance

Revenues

Concession contracts should lead to important financial gains for the government. One of the prevailing opinions is that a concession approach creates possibilities to realize projects that would not be carried out with traditional procurement because of budgetary deficit. A construction procured as a concession contract has to be paid for in several stages based on its life cycle costs, whereas traditional projects have to be paid for immediately after realization. This could create extra financial space for a government. However, according to many governments, a budgetary deficit should not be the motive for starting a concession. At the same time, we note that, historically, concessions are politically admired in times of economic recession.

It is generally agreed that the public sector can borrow more cheaply than the private sector; Ball et al. (2000) give two reasons: the size of the public sector and its consequent ability to bear risks, together with its tax raising powers. They concluded that the current weighted average cost of private sector capital on concession projects was between one and three percentage points higher than public sector borrowing. However, these extra costs are counterbalanced by important efficiency gains during construction and operation.

The fact that the concessionaire is also responsible for the project in the period after its realization adds value to both public and private parties. Public parties will obtain more financial security, while the concession contract generates continuity for private parties. Besides, it is more attractive for private parties to become involved in a project from the development phase onwards and thus be involved in different aspects of the project, than it is to have to calculate costs on the basis of a specified design, as is the case in more traditional projects.

To the best of our knowledge, there is no broad evaluation study on the revenues in concessions. Allowable rates of return range from 15 to 25%

depending on the level of risks involved. Kumaraswany and Morris (2002) found projected rates of return in early projects in Hong Kong to be 15 or 16%. For the Pakistani power projects, Malaysian water supply project and the Bangkok Second Stage Expressway, they found rates of 16%, 18–20% and 21%, respectively.

The revenues for the various partners in a private consortium depend on the specific contract structure within the consortium. Most contracts are turnkey contracts. The shareholders get a percentage of the revenues equivalent to the percentage of the work they are taking. The contract normally regulates when and on what conditions partners can withdraw. In some concession contracts, the consortium is able to exploit not only the infrastructure but also the commercial spaces within the statutes; this is an example of external added value (see Section 2.2.1). In most cases, the concession contract gives no guarantee that the consortium will also win new contracts for new railways.

Efficiency gains

Increased efficiency is an important motive for governments to launch concession projects, these having been shown to be superior when compared to a hypothetical traditional approach using a public sector comparator; for example, a highway can be constructed more quickly using a concession approach than by means of a traditional approach. However, it remains uncertain what would have happened in the traditional scenario.

It is argued that the concession holder will be more efficient than their public counterpart because the concessionaire will be penalized for cost and time overruns, although in practice, concession holders will be only penalized if the public agent can show that time and cost overruns are the fault of the concession holder.

In their evaluation study of concession contracts the National Audit Office (NAO) (2003) concluded that in the UK concessions brought significant benefits to the government in terms of delivery of built assets and the price expected by the public. Seventy-six per cent of the projects they surveyed were ready to use on time or earlier. Moreover, if time and costs were exceeded in most projects this was caused by changes required by the client. Reasons for the price increases were new facilities, extensions and enhancements to facilities. However, the NAO clearly indicated that it was not possible to judge what the effects would have been when using different procurement schemes; in general, there is little empirical evidence that the private sector can operate more efficiently than the public sector (Hall, 1998).

In contrast to the NAO study, Ahadzi & Bowles (2004) reported on various studies showing excessive time and cost overruns. These overruns were found during the pre-contract stages and resulted in huge advisory costs. Allen (2001) even found cost overruns of up to 600%. According to Ahadzi and Bowles (2004) the contract negotiation phase is the most critical stage during which delays are most prominent.

Since many concession contracts include transfer of personnel, important cost savings can also be attained by reduction of staff or changing employment conditions (Ball et al., 2000). When the contract coincides with the transfer of employees from the public to the private sector, strong opposition often results. For instance, cleaners in privately financed hospitals work under conditions different from publicly employed personnel. Unions in the UK see concessions as a threat to their members because they have to operate in a more commercial way. However, according to private parties, concession contracts stimulate employees to work more efficiently.

Another reason that a concession leads to greater efficiency is that they enable the government to focus on strategic and policy-making services, instead of the execution focus that has prevailed until recently.

In summary, several arguments can be formulated to indicate that concessions may lead to less time and cost overruns, but studies have shown that concession arrangements are no guarantee of efficiency. As Allen (2001) concluded, 'cost overruns can and do occur under both private sector and public sector management'.

Transaction costs

It is generally argued that concession PPP projects should have certain economies of scale. Projects need a specific volume (in terms of concession period and costs) to recover the transaction costs. Bidding costs are high. Because of these high costs in the UK, often only a few bidders sign up. The bidding costs for a hospital, for instance, may amount to about €4.5–6 million. In many smaller countries, as is the case in The Netherlands, concession PPP projects are relatively small and therefore time- and cost-consuming. According to many private actors in The Netherlands, the high speed train was a fair project to start with because it was large enough to cover the research and development costs. Moreover, for every project a new contract has to be developed and negotiated which is time-consuming.

High transaction costs may be a barrier for private companies to take part in the bidding process. The Anglo-Saxon bidding process is not only time-consuming; the complexity of contract development leads to high legal fees. In comparison to traditionally procured projects, the legal fees are estimated to be up to 20% higher (Construction Industry Council, 2000). Exact costs for tendering are hard to find because they are considered to be confidential. Allen (2001) concluded that the total costs for tendering for concession contracts were under 3% of expected total costs, while for traditional procurement total tendering costs were under 1%.

The relatively large, long term or sequential projects (smaller projects which are executed together) are therefore considered more suitable for concession PPP procurement. Life cycle value becomes important. Because the payments are spread over the concession period, the concessionaire will be encouraged to

Table 4.5 French concession risk allocation strategy (US Department of
Transportation, 2004).

Revenue and traffic risk	Concession
Construction cost risk	Concession
Financial risk	Concession
Operation cost risk	Concession
Project risk	French state
Force majeure	French state
Government action	French state

minimize the 'lifetime costs' by balancing higher construction costs against lower
maintenance costs in the future (Hall, 1998). Moreover, payments based on the
availability of the service gives an incentive for optimal maintenance.

Risks

One facet of a concession is the possibility of allocating risks to both principal and
concessionaire. Efficient allocation of risks is one of the key issues of concessions.
According to Hall (1998) there are two key questions concerning risks: *how much*
risk and *what type* of risk is transferred. These questions should be addressed
during the development of the business case, when benefits, requirements and
scope of works are defined. Overestimating the revenues or underestimating costs
and time cause optimism bias that can be expressed as the percentage difference
between the estimate at appraisal and the final return (Mott MacDonald, 2002);
according to their evaluation, optimism bias is primarily caused by the failure to
identify and manage project risks.

In practice, defining the exact allocation of risk is difficult and complex. Risks
should be transferred to the party that is best able to manage them. In France, a
risk allocation strategy is defined for concessions, as is shown in Table 4.5.

To give another example, in Spain the partners are covered for possible losses
if the contract changes or the government wants to interfere. In the contract it is
mentioned that, if the government promotes a competing system, the partners
will be compensated. In the contract the risks are allocated to the company/part-
ner in the consortium that can manage the risks best.

Hall (1998) mentions two methods of risk transfer that lead to poor value for
money because the risks are largely outside the control of the private operator:
volume risk and residual value risk:

- *Volume risk* The principal payment mechanism under the contract is related
  to volume rather availability. The client controls the volume, not the private
  operator, and therefore these risks should be taken by the client. In the first
  concession, road contracts payments depended on the volume of traffic and
  these were paid by the government, not by the users.

■ *Residual value risk* In many projects, payments are related to how much the assets will be worth at the end of the service contract. Whether the risks associated with the residual value can be controlled depends, according to Hall (1998), on whether the residual value is subject to decisions taken by the contractor, and this in turn will depend on whether a competitive market exists. In office accommodation, for instance, such a competitive market exists and hence private contractors can control some of the risks, but for many other assets it is unlikely that a competitive market will develop.

A survey among more than 120 projects in the UK (National Audit Office, 2001) indicated that the allocation of risks was either wholly or partially appropriate; however, only 50% of the private sector partners believed that the project risks had been allocated optimally in contrast to 80% of the public sector partners. A study by Mott MacDonald (2002) showed that estimates of revenues and risks (in their terms the optimism bias) were assessed more accurately in concession projects than in traditionally procured projects because in concessions the project risks are passed to the party that is best placed to manage them.

Besides identifying which risks could be transferred, the risk management process in concessions could be improved by professionalization and standardization. Private respondents in a survey undertaken by Akintoye et al. (2003) gave the following recommendations for improvement:

■ Improve the clients' expertise
■ Further standardization of the risk assessment and management
■ Develop a national database for historical records. (This recommendation could be seen as a plea for transparency in the costs systems of the organizations involved.)

These recommendations indicate how important a national knowledge centre can be.

Content performance: innovation

Innovation is often mentioned as one of the benefits of concessions in the political debate and in the literature, but mostly in relation to financial indicators. Third party income, it is argued, is favoured as public sector managers may not encounter incentives to take risks through innovation (Hall, 1998). The opportunity for innovation is defined in terms of funding packages, delivery of services and construction of the asset (Ball et al., 2000), i.e. innovation leads to savings in construction and operation costs. Innovation is a tool for cost savings, not a goal in itself. Nevertheless, innovation is possible because the client produces an output specification instead of a type of service.

In a workshop in 2004 with both public and private decision makers involved in concession projects in The Netherlands, it was concluded that technical innovation was not and is not an important motive for starting concession projects (Dewulf et al., 2004). The participants in the workshop were convinced that concession projects would not lead to more technical innovation. The financial component of the project should not be allocated to the concessionaire, which is the case in concession projects, when technical innovation is wanted. Neither does the fact that financial institutions are involved in concessions contribute to more innovation, because these institutions avoid uncertainties. Apparently, financial security within concession projects is too important to include innovative (and therefore uncertain) aspects in the project.

Some case evaluations indicate that concession arrangements have not resulted in innovation. An important reason for this finding is given by Ball et al. (2000) in their evaluation of a high school project. The authors found that the private sector was advised, before bid submission, of the 'acceptability or otherwise of particular design solutions' through informal meetings with the local authority. This result agrees with the statement a senior public servant in the UK made to us. According to this officer, concession contracts do not result in innovation because of the following:

- Government people are very cautious in promoting innovation or alternatives
- The private sector is risk averse, being convinced that if they want to win the bid they had better stay close to the original design.

Other cases have shown that concessions may lead to technological innovation. In the case of the Tate's Cairn Tunnel, ingenious engineering solutions were developed in tunnelling and tube construction which led to considerable savings in construction time (Kumaraswany & Morris, 2002). Also in this case, innovation was not the goal but the means by which to construct more efficiently.

An important barrier in many countries for design innovation is that of legal procedures. In The Netherlands, for instance, design innovation is difficult due to the reference design, which complies with the Tracébesluit that regulates the moment of private involvement in the decision-making process. Deviation from this reference design is very time-consuming and expensive, and for this reason not favoured by the government. If the private partners could have been involved earlier in the process, they would have been able to draw up the design to be used for the Tracébesluit, which they believed would have resulted in important cost savings.

Many scholars believe that to ensure meaningful stakeholder involvement, it must occur 'early, often and ongoing' (Wondolleck & Yaffee, 2000). Early interaction injects community knowledge and expertise into the planning process when it is most needed, before politics are set in stone (Graaf, 2005). However, early involvement is contrary to European procurement rules which stress the importance of transparency of selection. Early involvement could lead to exclusion of a private party from bidding.

In many tendering processes, innovation is not a criterion for selection. Selection takes place on the lowest bid. Innovative solutions are welcomed, but are discussed during the negotiation phase (after selection) of the tendering procedure as is the case in, for instance, Portugal.

Besides the performance criteria described in Chapter 2, we can reveal an important side effect of the concession policy. In several countries, market organizations have become world leaders in construction because of the large concession projects launched by their governments. In Australia, for instance, this was recently stressed by Crump & Slee (2005) who report that the Victorians are paying twice as much for PPP tollways than if they had been financed by government borrowing, yet 'the community is asked to feel gratitude to the tollway company for building the road'.

4.5.2 Process performance

Despite the fact that many bottlenecks occur during the functioning of concessions, few empirical studies are available on the performance of the process. In many concession projects, renegotiations commenced shortly after the contract was signed due to changes in traffic flows, environmental issues or other changes in the context of the project. Because of these dynamics the public interest changes over time. In addition, many sociological factors have a strong impact on the performance of the concession, such as trust, perseverance and flexibility.

Actors' fit and willingness to co-operate

During the negotiation process discussions focus on price and risks. The public agent sets the output specifications; in most cases these specifications are not discussed, let alone altered. One could argue that in concession there is no true co-operation, rather a clear separation of tasks and responsibilities between the client and contractor. In contrast to joint venture PPPs, consensus of goals is not needed to start a concession project. However, understanding each other's goals and interests is a condition for success. In many cases the requirements of the public client are not well understood by the concession holder. Hence, in the UK, PFI projects should start by using the design quality indicator (DQI). This tool was developed to generate consensus on design evaluation and helps the procurement team to define and check the design quality beforehand and during the building project. In 2007, 60% of all PFI projects should be using this DQI tool (Gann & Whyte, 2003).

Public interest

Critics of private finance contracts often argue that public interests are at stake. The debate on public interest is vague and political. What is the public interest in spatial or infrastructure projects and how can the government safeguard this interest in the force field of parties that are involved in these projects? Despite the attention given to the concept of public interest in the media and in public debate, few empirical studies are available and the research is mainly limited to the description of developments in specific cases and selected aspects. With regard to these aspects, the focus is primarily on enhancing financial efficacy instead of paying attention to the best way to safeguard the public interest (Smit & Dewulf, 2004).

Behaviour

There is little empirical evidence on the role of behavioural aspects in concessions. In some cases, as will be shown in Chapter 5, behavioural aspects do play an important role in the functioning of concessions. Because of the many uncertainties in the context of the project, flexibility is needed, as are trust and respect. Public and private actors stress the importance of these factors (Spiering & Dewulf, 2001; Dewulf et al., 2004), but in the literature little or no attention is paid to these issues. Trust, respect and flexibility form the basis of the portfolio contracts, as will be discussed in Chapter 8.

4.6 Lessons

4.6.1 Conditions for success

When discussing the performance of concession projects it became clear from the historical description that many factors influence their success. These factors differ strongly from country to country and it is therefore difficult to generalize. Success does not depend on a single factor but on a mix of conditions. In Chapter 2 it was argued that the characteristics of the actors involved in the projects, their environment or networks, and the specific characteristics of the project influence the functioning of PPPs in general. This distinction shows many similarities with the factors found for concessions by Ahadzi and Bowles (2004) who distinguish the characteristics of the public and private sector parties to the negotiations, the external environment, the organizational structure, strategies and culture, and specific project characteristics. Our findings correspond with these distinctions and we conclude that the success of concessions will depend on the factors listed below.

Context situation

Institutional conditions (legal aspects, e.g. procurement rules, procedures)
Many governments have adapted their legislation to develop optimal conditions for concession arrangements. In most legislative frameworks, the assignment of responsibilities for the delivery of public services is complex and restrictive. In addition, financial control mechanisms within the public sector do not allude

to public services being financed and delivered by the private sector. As a result, many countries have introduced new legislation to facilitate the delivery of public services by the private sector. It appears that the following aspects need attention:

■ Establishing or clarifying the legitimacy and powers of public authorities to enter concession contracts
■ Removing tax anomalies that can weigh against concession approaches
■ Refining public expenditure capital control regimes to accommodate concessions.

Market relationships (the relationship between government and the marketplace, as well as between construction firms and financial institutions)

An important issue raised by many authors is the level of competition. In our analysis we have mentioned several times that concessions could only be successful if a really open competitive market exists; in the words of Grimshaw et al. (2002, p. 479), 'competition is the main driver for ensuring continuous improvements in public service delivery'.

Cultural–historical situation

Several countries have a long history of concession contracts; others do not. Experience provides expertise through learning (Ahadzi & Bowles, 2004), so to become experienced a flow of concession projects should be developed.

Organizational variables

■ Size of the construction firms and financial investors. One of the great advantages of the Spanish market is the large size of the construction firms: scale is needed to be able to take substantial risks
■ Skills and capabilities of government and marketplace. An important issue in the historical description was the existence of a central PPP unit. According to Ahadzi & Bowles (2004, p. 976), the absence of such a unit may even raise concerns about the public sector's project management skills. The formation of PFI or PPP units, a centralised support unit, a public sector advisory group or a special task force, for example, is the best way to manifest governmental commitment. Within these units, a consistent and co-ordinated approach to concessions, policy guidance and standardized tendering documentation may be developed. Furthermore, these units can assist ministries to select, prioritize, scope and procure projects, and recycle knowledge and experience within and between ministries. The establishment of dedicated PPP units and task forces all over Europe has mobilized public administration behind concessions for high-level political sponsorship. It also appeared to be important that a minister would champion and promote the concession process within government.

A stable and confident government policy is also required. The enormous flow of concession projects in Spain and the UK is mainly the result of straightforward government policy. The government as a sponsor also has the responsibility of creating an adequate regulatory environment for concessions to develop (Romeiros de Lemos, 2002). One of the major pitfalls in concessions is the issue of risk aversion and management. The political agenda is one major risk. Today, concessions receive much attention from both the public and private sectors. Negative stories in the media, however, could lead to a drop in the number of concessions or to public opposition. One example of the impact of politics on the development of PFIs is found in Portugal. A change in the Portuguese administration led to renegotiations of the existing concession contracts because of the political desire to change the term of the contract from long term (say 30 years) to short term (say 5–6 years) and the political push to introduce tollways instead of shadow toll roads because of the large claim of the latter on the public purse.

Specific project variables

- Process approach
- Organizational structure
- Legal and financial framework.

These aspects will be further elaborated in Chapter 5, where we will discuss some projects in detail.

4.6.2 Improvements

As mentioned before, the performance of concessions depends on a wide variety of factors. Today, many governments are working on guidelines, standardization and professionalization, to improve the performance of concessions. Some of these improvements focus on the creation and others on the functioning of concessions.

In Chapter 5, we will elaborate further on improvements at specific project level and in Chapter 8 discuss some new approaches in PPPs that aim to cope with the shortcomings described. To end this chapter, we distinguish the following points for improvement:

(1)   Clients are too focused on details rather than on strategic issues: because of the relatively long duration of the contract, there is much uncertainty about the circumstances which might occur during the partnership. In general, the government feels inclined to allocate all the risks regarding these uncertainties via the contract. However, this makes the contract too detailed and therefore inflexible; furthermore, this type of risk allocation is not a good basis upon which to develop trust between the different actors in the contract.

(2) Transaction costs should be lowered: an important issue today is how bidding and management costs can be decreased. Standardization and shortening of the bidding process are two major improvements that governments work on continuously.

(3) Clients find it difficult to define clear performance or output criteria: many private partners involved in our research complained about the opacity of performance criteria and the way different bids are judged. This was especially the case in countries with long and complex procurement systems, such as the UK and The Netherlands.

(4) The quality criterion is inferior to financial criteria in the current way of judging: selection on price is a clear and transparent selection criterion. However, reform programmes in various countries worldwide were launched to generate more value for the client. One major aim of the Dutch reform programme in construction was to progress from a cost-orientated focus towards a value-orientated focus.

(5) Insufficient design innovation: although perhaps not the prime motive to start concessions, innovation in design becomes increasingly important because space is scarce and society asks for innovative solutions. Stimulating innovation and evaluating bids on the level of innovation will become an intrinsic part of the bidding process.

(6) The preferred bidder is often selected too late in the process: the private sector finds early involvement of greater importance than their public counterparts (Bult-Spiering & Dewulf, 2002; Ahadzi & Bowles, 2004). Early involvement generates, it is argued, more opportunities to influence the design and therefore to make an optimal proposal.

(7) The public–public alignment is often a burden for success: in many cases we see difficulties in alignment between national, regional and local levels. While executing the concession projects, problems arise because of the improper division of tasks between central and local governments. It is preferable to address one co-ordination unit to attune the different (governmental) departments and levels.

(8) A clear and well-structured methodology for choosing the right procurement strategy is missing: such methodology should facilitate choice of the way to procure.

(9) Comparative method (PSC) is not reliable: in traditionally procured projects, the price comprises nominal amounts, whereas net cash values are used in concession projects. Furthermore, in traditionally procured projects the risks are only partly incorporated, whereas in concession projects all risks are transparent. It could be said that the two estimates are incomparable.

(10) Specific concession knowledge is missing from the governmental organization: the governmental organization did not gain maximum benefit from former concession projects because external experts were hired and the knowledge gained about the concession projects disappeared when they did.

(11)   Risk allocation: risks may be spread over different small projects or a portfolio of projects, provided that the required economies of scale can be achieved.

4.7 Summary

In this chapter we have described the motives of various countries for implementing concession PPPs.

One aspect of concession PPPs is that financial arrangements are integrated. Consequently, several disciplines are brought together to generate added value for both principal and concessionaire. Financial clarity and transparency for the client organization as well as other organizations are required to come to an optimal division of finance and risks.

In several studies the motives for starting concession PPPs are mentioned. One of these motives is to encourage innovation. It appears, however, that (technical) innovation is not achieved by using a concession approach in construction projects. Because financial responsibility is allocated to the concessionaire and financial institutions are involved in concession projects, uncertainties are avoided, resulting in little innovation.

Transparency and trust in procedures are considered to be very important; at the same time, trust is one of the most elusive concepts. However, a shift from a system based on mistrust to a system based on trust is necessary if optimal partnering and co-operation processes in concession PPPs are to be achieved.

This chapter has further described the various types of concessions. The concession procedures and procurement approaches differ from country to country. The approaches are to a certain extent based on historical developments and are partly the result of different motives. In Chapter 5 the characteristics of the various approaches will be illustrated by examples.

References

Ahadzi, M. & Bowles, G. (2004) Public–private partnerships and contract negotiations: an empirical study. *Construction Management and Economics*, **22**, 967–978.

Akintoye, A., Hardcastle, C., Beck, M., Chinyio, E. & Asenova, D. (2003) Achieving best value in private finance initiative project procurement. *Construction Management and Economics*, **21**, 461–470.

Allen, G. (2001) *The Private Finance Initiative*. Research paper 01/117, Economic Policy and Statistics section; House of Commons Library.

Association of Chartered Certified Accountants (ACCA) (2004) *Evaluating the Operation of PFI Roads and Hospitals*. Research report no. 84. London: Certified Accountants Educational Trust.

Austin, J. & McCaffrey, A. (2002) Business leadership coalitions and public–private partnerships in American cities: a business perspective on regime theory. *Journal of Urban Affairs*, **24** (1), 35–54.

Ball, R., Heafey, M. & King, D. (2000) Private finance initiative – a good deal for the public purse or a drain on future generations? *Policy Press*, **29** (1), 95–108.

Bates, M. (1997) *First review of the private finance initiative*. London: HMSO.

Beierle, T.C. (2002) The quality of stakeholder-based decisions. *Risk Analysis*, **22** (4), 739–749.

Bongenaar, A. (2001) *Corporate Governance and Public Private Partnership: The Case of Japan*. Utrecht: Nederlandse Geografische Studies.

Brody, S.D., Goschalk, D. & Burby, R. (2003) Mandating citizen participation in plan making. *Journal of the American Planning Association*, **69** (3), 245–263.

Bult-Spiering, W.D. & Dewulf, G.P.M.R. (2002) *P3BI in PPS: een Visie*. Enschede: P3BI.

Chang, M. & Imura, H. (2003) Developing private finance initiatives (PFI)/public–private partnerships (PPP) for urban environmental infrastructure in Asia. *First Task Force Meeting on Financial Mechanisms for Environmental Protection*, Institute for Global Environmental Strategies, Graduate School, Nagoya University.

Construction Industry Council (2000) *The Role of Cost Saving and Innovation in PFI Projects*. London: Thomas Telford.

Committee of Public Accounts (2003) *Delivering better value for money from the private finance initiative*, Committee of Public Accounts, 28th report. London: House of Commons.

Crump, S. & Slee, R. (2005) Robbing the public to pay private? Two case studies of refinancing education infrastructure in Australia. *Journal of Education Policy*, **20** (2), 243–258.

Dewulf, G., Bult-Spiering, M. & Blanken, A. (2004) *Opportunities for PFI in The Netherlands*. Enschede: P3BI.

Elbing, C. & Jan, A. (2005) Public private partnerships in Europe – track to success, bottlenecks and lessons learned from different countries. Paper presented at *12th Annual Conference of the European Real Estates Society*, Dublin, Ireland.

Forester, J. (1989) *Planning in the Face of Power*. Berkeley: University of California Press.

Gann, D. & Whyte, J. (2003) Special issue on design quality indicator. *Building Research and Information*, **31** (5).

Graaf, R. de (2005) *Strategic urban planning: industrial development in The Netherlands, to direct or to interact?* Dissertation. Enschede: University of Twente.

Grimshaw, D., Vincent, S. & Wilmott, H. (2002) Going privately: partnership and outsourcing in UK public services, *Public Administration*, **80** (3), 475–503.

National Audit Office (2001) *Managing the relationship to secure a successful partnership in PFI projects*. HC375 Session 2001/2002, 29, November. London: National Audit Office.

Guislain, P. & Kerf, M. (1995) *Concessions – the way to privatize infrastructure sector monopolies*, Public Policy for the Private Sector, note 59. Washington DC: World Bank.

Hall, J. (1998) Private opportunity, public benefit? *Fiscal Studies*, **19** (2), 121–140.

HM Treasury (2003) *PFI: Meeting the investment challenge*. London: HMSO.

Innes, J.E. (1998) Information in communicative planning. *American Planning Association Journal*, **64** (1), 52–63.

Jacob, D. & Kochendörfer, B. (2000) *Private Finanzierung öffentlicher Bauinvestitionen – ein EU-Vergleich*. Berlin: Ernst & Sohn.

Koppenjan, J. (2005) The formation of public–private partnerships: lessons from nine transport infrastructure projects in The Netherlands. *Public Administration*, **83** (1), 135–157.

Kumaraswany, M. & Morris, D. (2002) Build–operate–transfer-type procurement in Asian megaprojects. *Journal of Construction Engineering and Management*, **128** (2), 93–102.

Li Bing & Akintoye, A. (2003) An overview of public–private partnership. In: A. Akintoye, M. Beck & C. Hardcastle (Eds) *Public-Private Partnerships: Managing Risks and Opportunities* (pp. 3–30). Oxford: Blackwell Science.

Mott MacDonald (2002) *Review of Large Public Procurement in the UK*, July, Treasury paper 200505. Croydon: Mott MacDonald.

Mustafa, A. (1999) Public–private partnership: an alternative institutional model for implementing the private financial initiative in the provision of transport infrastructure. *Journal of Project Finance*, **5** (2), 64–79.

National Audit Office (2003) *PFI: Construction Performance*. London: National Audit Office.

Norment, R. (2002) *The Changing Environment for Public-Private Partnerships in the United States*. Washington DC: National Council for Public–Private Partnerships.

PPP Knowledge Centre (2002) *Manual Public Private Comparator*. The Hague: Ministry of Finance.

PPP Knowledge Centre (2003) *Rijksbetrokkenheid bij Integrale Gebiedsontwikkeling en PPS*. The Hague: Ministry of Finance.

Regieraad Bouw (2004) *Van Raad Naar Daad*. Gouda: Regieraad Bouw.

Romeiros de Lemos, M. (2002) *Sustainable Competitive Advantage in PFI: A Systematic and Holistic Approach to Identify the CSFs in Risk Management in PFI, Taking into Account the Whole Life Cycle*. Salford: University of Salford.

Scott, S. & Harris, R.A. (2004) United Kingdom construction claims: view of professionals. *Journal of Construction Engineering and Management*, **130** (5), 734–741.

Smit, M. & Dewulf, G. (2002) Public sector involvement: a comparison between the role of the government in private finance initiatives (PFI) and public private partnerships (PPP) in spatial development projects. In: *Public and Private Sector Partnerships: Exploring Co-operation* (pp. 451–463). Sheffield: Sheffield Hallam University.

Smit, M. & Dewulf, G. (2004) The public interest in spatial (re)development projects. Paper presented at the *Conference of the Institute for Government Studies*, University of Twente, Enschede.

Spiering, W.D. & Dewulf, G.P.M.R. (2001) *Publiek–Private Samenwerking bij Infrastructurele en Stedelijke Projecten*. Enschede: P3BI.

Tiong, R. & Anderson, J. (2003) Public–private partnership risk assessment and management process: the Asian dimension. In: A. Akintoye, M. Beck & C. Hardcastle (Eds) *Public–Private Partnerships: Managing Risks and Opportunities* (pp. 225–244). Oxford: Blackwell Science.

Torres, L. & Pina, V. (2001) Public–private partnerships and private finance initiatives in the EU and Spanish local governments. *European Accounting Review*, **10** (3), 601–619.

US Department of Transportation (2002) *Contract Administration: Technology and Practice in Europe*. Washington DC: US Department of Transportation, International Technology Program.

US Department of Transportation (2004) *Report to Congress on Public–Private Partnerships*. Washington DC: US Department of Transportation.

Wondolleck, J. & S. Yaffee (2000) *Making Collaboration Work*. Washington DC: Island Press.

Further Reading

Akintoye, A., Beck, M. & Hardcastle, C. (Eds) (2003) *Public–Private Partnerships: Managing Risks and Opportunities*. Oxford: Blackwell Science.

Bult-Spiering, M. (2003) *Publiek–Private Samenwerking: de Interactie Centraal*. Utrecht: Lemma.

Cochran, C.E. (1974) Political science and 'the public interest'. *Journal of Politics*, **36** (2), 327–355.

Flanagan, R. & Jewell, C. (2005) *Whole Life Appraisal for Construction*. Oxford: Blackwell Science.

General Accounting Office (GAO) (2004) *Highways and Transit: Private Sector Sponsorship of and Investment in Major Projects Has Been Limited*. Washington DC: US GAO, GAO–04–419.

General Services Administration Office of Governmentwide Policy (2003) *Best Practices in Real Property Management in State Governments*. Washington DC: General Services Administration.

Hoogendijk, F.A. (2000) *De Verzorgingsstaat: Institutionele Vernieuwing en Publieke Belangen*. Den Haag: Ministerie van Economische Zaken.

McDowall, E. (2003) Applications of risk management strategies in public–private partnership procurement. In: A. Akintoye, M. Beck & C. Hardcastle (Eds) *Public–Private Partnerships: Managing Risks and Opportunities* (pp. 183–204). Oxford: Blackwell Science.

Ministry of Economic Affairs (2000) *Publieke Belangen en Marktordening: Liberalisering en Privatisering in Netwerksectoren*. Den Haag: Ministry of Economic Affairs.

PSIB (2004) *Inventory of International Reforms in Building and Construction*. Gouda: PSIB.

Raad voor het Openbaar Bestuur (1998) *De Overheid de Markt in- of Uitprijzen*. Den Haag: Raad voor het Openbaar Bestuur.

Webliography

Federal Highway Administration international.fhwa.dot.gov (accessed July 2005)

HM Treasury www.hm-treasury.gov.uk (accessed July 2005)

National Council for Public–Private Partnerships www.ncppp.org (accessed July 2005)

United Kingdom Parliament www.publications.parliament.uk (accessed July 2005)

5 Examples of Concession Projects in Europe

Chapter 4 described the development of concession policy in different parts of the world and the rationale behind this policy together with the added value. In this chapter we will highlight some projects to illustrate the characteristics described in the previous chapter. These examples must be seen as snapshots in the history of concessions. We focused on the creation of the concession arrangements to illustrate the way these concessions were procured and the criteria used for selection of the consortium.

5.1 European policy

The construction market in Europe is still a local one. We see major differences in planning and procurement between the European countries, with quite large differences between regions within countries. Spain, for instance, can be characterized by a federal system with 17 autonomous regions called comunidades. For historical reasons, some comunidades, such as Euskadi (the Basque Country) and Catalonia, have more competencies than other regions. In Germany, the Bundesländer have high levels of autonomy. Consequently, we see major differences in concession policy among European countries and regions. Some regions stimulate private financing, others not. In Germany, for instance, a concentration of concession schemes can be found in Nord-Rhein Westfalen. In Denmark, most of the concession schemes can be found at local government level; despite the central acceptance of public–private partnership schemes, concessions have not thrived at central government level (Greve, 2003).

In the past few decades, national and European governments have been working towards uniformity and standardization. In most countries, central legislation has been implemented and central knowledge centres or task forces introduced to stimulate uniformity in concession policy. One of the major activities of the national centres was, and still is, to develop standard guidelines and procurement models for concessions.

Furthermore, as a result of globalization of the market, the European Community is stimulating uniformity in concession policy between member states. For more than 40 years, European countries have been working towards unity. The EU is now a free market with a single currency; this needs standardization of concession policy. The recently published Green Paper (Commission of the European Communities, 2004) is a big step towards a standard approach. Uniformity leads to transparency and fair competition.

Concession contracts have many areas of application. Examples can be found in hard and soft infrastructure. The applications differ by country. Italy, France and Spain have used concession contracts for building motorways for many years, whereas Australia, New Zealand and the USA have been implementing concessions for the construction and maintenance of prisons, roads and hospital buildings.

Many examples of concessions can be found in schools, infrastructure and health care. We shall describe three different cases from three different areas of applications and countries: a school in The Netherlands, an underground railway in Spain, and a hospital in the UK. The cases differ in approach and content, and therefore provide good examples of the broad variety of concession schemes in Europe.

5.2 Concessions in schools: the case of Montaigne Lyceum[1]

Many European governments have announced major investment programmes to renovate or refurbish schools. Despite ambitious plans, concession in education has been slow to get off the ground but is now developing a rapid pace in countries such as the UK and Germany. For example, in the UK, more than 500 primary and secondary schools were part of concession deals signed or in procurement during 1997–2004 (Audit Commission, 2003). Concession policy in education covers a broad variety of partnerships between public and private actors, whereby a private contractor provides the facility and takes on its long term operation in line with the specification set out by the public agency (Edwards & Shaoul, 2003). The public agent then pays a fee for the use of the assets together with a facility management fee for services.

An interesting case from The Netherlands that illustrates the way concession schemes are structured in education in many western countries is provided by the Montaigne Lyceum in The Hague. This secondary school will offer vocational, pre-university and general secondary education in Ypenburg, an ambitious urban housing project in The Hague. The project encompasses:

- 1200 pupils
- 10 000 m² of buildings
- 30 year contract
- buildings, furniture, information and communication technology infrastructure, energy, cleaning and removal.

---

[1] Information about this case was gathered from interviews with the Ministry of Education, Strukton and PPP Knowledge Centre. More information can be found on the PPP Knowledge Centre website www.pps.minfin.nl/.

In the Netherlands, the budgets for the construction and major maintenance of schools are transferred to the city councils, while the school boards themselves receive a compensation for minor maintenance and facilities once a year. The DBFM approach is only possible when agreements are made about the integration of these cash flows.

The school board and the local government foresaw several efficiency gains from the introduction of a concession scheme. Several advantages were expected: a better price/quality ratio (reduction in maintenance and exploitation costs in combination with quality improvement) and the fact that the teaching staff could direct their attention to education. Additionally, the innovative procurement was seen as an example of the mission of the school: innovative, creative and challenging (Andersson Elffers Felix, 2005). The concessionaire should accept responsibility for the availability of qualitatively good education, including the day operation of the school in terms of security, technical installations and maintenance, ICT infrastructure, cleaning, catering and so on. The municipality commissioned the project because of financial inducements. However, in practice most activities in the tendering phase were executed by the school board and sanctioned by the city council. The Dutch knowledge centre on PPPs as part of the Ministry of Finance is also involved in the project to standardize contracts so that 75% of the contract conditions will be standard conditions in the future. This is a pilot project of the Dutch Ministry of Finance on the basis of which the Dutch government wishes to study the opportunities for launching concessions in education.

The total planning phase of the project took 18 months. The time taken to reach an administrative arrangement, including project-specific agreements between the municipality and the school board, was particularly long.

Despite the project being relatively small, 11 proposals were submitted in the first round. From this list, a shortlist of four bidders was selected and negotiations went ahead. Finally, the consortium TalentGroup won the deal: TalentGroup comprises Strukton, Barclays Private Equity, Imtech and ISS Facility Services. The Montaigne Lyceum should be ready for use in the summer of 2006.

The volume development of a school is difficult to estimate: therefore the risks with respect to ranging load factors were allocated to the school board because private organizations can neither influence these risks themselves, nor estimate this volume better than the government. Besides, demographic development risks, unlike foreseen developments, are typically risks to be reserved for the public domain.

Because of the small scale of this project, the benefits of the concession approach are relatively limited in terms of the comparatively high transaction costs and initial expenses of the first privately financed project in this field. As the result of administrative scaling-up in the educational sector, a portfolio of similar projects could be a realistic scenario for the near future, which would make concession approaches in this sector more attractive.

The DBFM contract is based on value for money methodology. Moreover, a building was procured together with delivery of accommodation services. The bidders were encouraged to deliver alternative and smart solutions for the output specifications defined by the government. In other words, the private consortium received more freedom than with traditional procurement to develop new alternative designs. To encourage the consortium to deliver value for money several measures were undertaken. The payment, for instance, will depend on the performance of the service to be delivered; an 'availability fee' will be paid by the public authority. The building has to be available for educational purposes. Availability could depend on, for example, the building climate and the light. The contract between the consortium and the school recognizes a long list of quality criteria or, in other words, output specifications. The contract, moreover, describes in detail what the consequences will be of non-delivery of the required level of service.

An important condition for the success of such a case is the competitive tension needed to deliver value for money. In the case of Montaigne, many private consortia were interested, presumably because it was expected that more concession schemes would be introduced in education in the near future.

5.3 Concessions in transport infrastructure: the case of the Metro de Sevilla[2]

Spain has a long tradition of concession contracts in transport infrastructure. In the 1960s, the government simply did not have the funds to build infrastructure; consequently, concession contracts were introduced. Large Spanish contractors, such as Dragados and Ferrovial, are today operating in many regions all over the world, especially in Latin America, Eastern Europe and South Africa. The experiences with concession contracts in Spain are now disseminated to the rest of the world. Although in Europe many countries are copying concession approaches from the UK, Spain could be seen as the cradle of concessions in the world and certainly the Spanish speaking part of the world. To ignore Spanish concession experiences would be a great omission in a book on PPPs.

Today, the motives to use concession contracts are primarily financial, as was the case for the early projects in the 1960s. With the exception of a short interval in the 1980s, the Spanish government has been promoting mixed public–private financing. Concessions have been seen as a necessity to cope with budget deficits. Both public and private parties in Spain stress the advantages of concession contracts in terms of transparency and flexibility. Concessions are seen as a way to operate more efficiently.

In addition to budget reasons, it is important to mention that in Spain it is widely accepted that the user of the infrastructure pays provided that an

---

[2] Information about this case is based primarily on documentation received from and interviews with the concessionaire Dragados.

alternative is available. In some other countries, for instance The Netherlands, for a long period it was politically unacceptable that civilians should pay for the use of infrastructure: roads and other forms of hard infrastructure were seen as a public good and hence the responsibility for provision lay with the government as guardian of the public interest; introduction of tollways would have been political suicide.

An interesting case that illustrates the way Spanish concession contracts are procured and managed is the Metro de Sevilla. Seville is the capital city of Andalucia and is the fourth largest metropolitan area in Spain. The plans for building a new underground railway network dated from the 1960s. The construction of the first line of a full three-line metro network started in 1978, but the project was suspended in 1983 for technical and financial reasons. It was first feared that historic buildings in the city centre might be damaged; a survey then concluded that the metro network would not be profitable because of insufficient patronage (EMTA News, 2003).

It took 15 years before a new initiative to develop the metro was launched. With metro construction booming in many other cities in Spain during the late 1990s, Seville and Andalucia rediscovered the need for improved public transport and started a plan for the construction of a light rail network in Seville. The new plan comprised four routes and 54 km of metro lines. A tendering procedure was developed for the building and operation of the first line comprising 19 km of track and 23 metro stations to link the south-eastern suburbs with the western part of the metropolitan area; a 35 year concession contract was issued.

In mid 2002, a request for proposals was published. The government developed the design. In the bid, there was little scope for adjusting the design: changes could be made to make it cheaper and feasible. Three months were provided during which to submit the bid and the process closed in November 2002. Three international consortia submitted bids which were studied by the client until March 2003. The proposals were similar in (technical) design but differed in financial terms (price). In March 2003 the contract was granted to a consortium comprising Dragados and Sacyr; it took one month to close the financial contract and sign the consortium agreement. After the contract was agreed upon by the consortium and the public authority, the partners had to negotiate with the European Investment Bank (EIB) regarding the specific financial details. Although the EIB had an agreement with the government before the bid was launched, there was a small chance that the EIB would back off.

Fifty per cent of the project is being financed by the EIB. It is doubly subsidized, thus:

- One third of the investments (except for the rolling stock)
- Subsidy of the tariff (bidders state how much they require).

Total investment in the project is €450 million, of which

- 20% is equity (€90 million)
- 33% is subsidy (33% (€450 m − €50 m rolling stock) = €132 million)
- €230 million is borrowed from the EIB.

Work began in the summer of 2003 and operation of the metro is planned for June 2006.

The Metro de Sevilla is a good example of the way in which tradition in Spain has influenced the procurement process. The bidding process has a restricted time-frame; there is no pre-qualification, preparation for tendering takes four months, and the award may take up to a maximum of four months. The costs for bidding are estimated to be approximately €0.5 million.

The awarding criteria that were used were clear. Points were allocated for specific technical (maintenance, level of service, electronic panel) and commercial criteria (including tariffs and subsidies) and used to weight the criteria in the tender as follows:

■ Technical offer: 500 points
■ Economic offer: 400 points
■ Integration with other transportation modes: 100 points
■ Total: 1000 points.

The Spanish government has only limited requirements when signing off a contract:

■ When the proposal is accepted and the contract signed, the concessionaire has one month during which to pay the government. Financial closure occurs after the bidding process. In contrast, the bidding process in the UK, Ireland and Portugal takes up to two years. The government in these countries asks all the financial partners for confirmation; external lawyers are needed to support this financial closure
■ There is no concession contract negotiation process. The concession contract is known by all bidders and remains unchanged from process to process. This is seen to be convenient for the bidders.

Before releasing a new tender, the following tasks are undertaken by the Spanish government:

■ Preliminary design preparation (1/5000 scale)
■ Environmental studies and required approvals
■ Public information period (in accordance with the Spanish legislation period). Public interest is guaranteed through involvement of the public by means of a survey through which they are asked about the required quality
■ Financial feasibility study and EIB consultation.

All of the above will facilitate the future work of both the government and the contractor after the process has been launched. Having all the client-related approvals obtained beforehand reduces risks and therefore prices.

During evaluation of the tender, bidders are given the opportunity to review each other's offers for several reasons:

■ Increase of market competitiveness: bidders can learn from what others have developed and included in their offers, and use it in subsequent processes
■ Complete review of the offers: any inconsistency in the offer can be discovered by the competitors and notified to the government
■ Complete transparency makes it possible to check clearly whether one offer is better than another one, and if the points have been awarded accordingly.

The case of the Metro de Sevilla also illustrates some pitfalls of the Spanish system. In Spain, the suppliers of rolling stock are not usually enthusiastic to participate in the concession; their core business and competence consists of delivering operational services. The client wants to involve the suppliers in the project, but they are not interested in the design and build phase. The provider and the operator want the contractor to take risks associated with the contract, but want to run the operation on their own. This problem always arises during negotiations. Often, after the bidding, the negotiation process ends badly. In the Metro de Sevilla project, one of the early bidding consortia had this problem, so only three competitors were left at the end. The government believes involvement of the contractors and integration of the disciplines to be important in order to obtain a good image for the city and excellent train performance at the end of the operation.

If the private partners had had to provide the design themselves, the bidding would have cost a great deal of money. Costs would have had to be charged to either the government or the public payer. Additional problems that could have arisen from early involvement of and designs by the private partners were the availability of land (expropriation) and the environmental impact. Innovation could take place through new construction techniques that may be submitted in the bid.

In our research, the concessionaire indicated some other pitfalls:

■ Integration of the system (infrastructure and operational services)
■ Risks in ensuring commitment between all parties involved in the contract (for foreign companies this is often even more difficult)
■ Difficulty in coping with violators (people who do not buy tickets, etc.)
■ Co-ordination between public systems; in particular, the tariff and time schedules have to be integrated.

Besides defining the criteria before the launch of the tendering process, the client also monitors performance during the contract period:

■ During construction
■ During the technical operation: at fixed milestones, the government determines the quality and imposes penalties in the event of shortcomings.

Bonuses could be awarded when specific performance criteria in the operating or maintenance stages (safety, waiting times, etc.) are satisfied. One bonus could be an expansion of the concession period by one year.

5.4 Concessions in hospitals: the case of West Middlesex University Hospital[3]

Western governments are confronted with the need for large investments in healthcare. Many current hospitals and other healthcare facilities are unsuitable for the provision of modern healthcare. To be able to deal with the growing financial needs, governments worldwide are considering the implementation of concessions in healthcare.

In the UK, government policy is that concessions have a central role to play in the capital programme for the National Health Service (NHS). In 2002, the Prime Minister stated that concessions should play a central role in modernizing NHS infrastructure; however, they are not seen as an alternative to, but rather as an addition to the public sector capital programme. According to the National Audit Office (2002) concessions have too often been considered as the only option. To justify the concession option, government departments have relied too heavily on the public sector comparator (PSC), which has often been used incorrectly as a pass-or-fail test. PSC outcomes have been applied with a spurious precision despite the many uncertainties involved in predicting future dynamics in healthcare. Moreover, numbers could have been manipulated to obtain the desired results. Before the concession route is chosen, departments need to examine all realistic alternatives and make a proper value for money assessment of the available choices. It is further stressed that matters other than purely financial considerations need to be appraised in assessing value for money (Association of Chartered Certified Accountants, 2004).

A good example of a concession scheme that proved to have value for money for the public sector is the case of West Middlesex hospital in London. This scheme comprises:

- 1200 staff
- 424 beds
- Ten operating theatres
- Two day-case theatres
- 155 000 outpatient attendances.

In 2001, the concession contract was let to the consortium ByWest. The contract was for 35 years and had a net present value of unitary payments of £125 million. After 35 years, the site will be returned to the hospital trust, but an extension to 60 years is possible. The contract required ByWest to redevelop the site in West London and to provide ongoing maintenance and facilities services. ByWest operates the building, and Liaison Group supervises the process. The organization consists of the NHS chairperson, chief executive and financial director, the

---

[3] Information for this case came from interviews with and documentation from the West Middlesex University Hospital, Department of Health and National Audit Office.

concession director, the ByWest chief executive consortium and the managing directors of the major parties in construction and facility management.

ByWest is a consortium with two shareholders: Bouygues and Ecovert. Bouygues is the design and construct contractor and Ecovert is responsible for facilities management. Abbey International is the primary funder.

In 2002, the National Audit Office assessed the concession contract for the redevelopment of West Middlesex University Hospital in positive terms. The scheme is being developed as part of the second wave of the PFI (or concession policy) in the NHS, in which a private sector organization is responsible for the construction and management of the buildings. The land remains the property of the NHS and only NHS doctors, nurses and therapists will provide patient care, as they do now. The West Middlesex hospital redevelopment is the most advanced of such schemes and is recognized for having pioneered a new form of concession contract. According to Sir John Brown, Head of the National Audit Office, the deal demonstrated that the NHS has learned lessons from earlier concession contracts. The hospital trust took account of the wider benefits expected from the concession deal (National Audit Office, 2002).

After evaluating the deal the National Audit Office (2002) came to the following conclusions:

- 'The deal meets expected local needs, with some flexibility to address inherent uncertainties in wider long-term NHS plans. Many of the buildings were over 100 years old and dilapidated. A redevelopment of the West Middlesex hospital site was essential to meet local needs for modern, high quality healthcare. One major problem is developing a concession scheme is that long-term planning is difficult because healthcare is changing over time and the local demography may also change. This may affect the optimum type and location of facilities that are required. For instance, the impact of more community-based services on the number of beds was and is hard to predict. This exposes the Trust to the risk that it may become locked into a long-term contract for buildings and services that are no longer needed. The NAO stresses that the long-term service contract of a concession deal makes termination likely to be more expensive. In the West Middlesex hospital deal there is some flexibility to accommodate these uncertainties. For instance, bed numbers could vary according to changes in demand. The Trust believes the contract provides sufficient flexibility to address future uncertainties in long-term healthcare.
- 'In getting the best available concession deal the Trust applied common sense and learnt from experience. The Trust ran an effective procurement and bidding competition based on own experience and new guidelines and standard contracts from the NHS. An important element was the faster bidding process. Extra rounds of bidding were eliminated which reduced the time and costs of both the Trust and bidders. ByWest's bid offered a slightly lower price than competing bidders.

- 'The Trust considered that the unquantifiable benefits of doing this as a concession deal outweighed the disadvantages. The financial comparison between the concession scheme and the costs of conventional procurement were not clear cut. But the Trust saw other benefits that would generate value for money. One such benefit was that the Trust believed that the contract would give the consortium incentives to complete the redevelopment quickly and to maintain the buildings well. A first evaluation showed that this was the case. In only 18 months the new hospital was constructed. Other important benefits were the price certainty and risk allocation to the private sector.'

The case of the West Middlesex hospital, moreover, shows that private companies are interested in getting involved in concession schemes in healthcare. There were 39 expressions of interest of which 13 consortia could be considered to be serious bidders.

The hospital trust ran an effective bidding process. Of the 13 consortia, six were put on a long list; three of these were short-listed and ByWest was finally selected as preferred provider. Reduction of time and costs was made possible because ByWest agreed, during the finalization of the contract, upon the price and other terms of the deal. Besides the lowest price, the bid showed the best value for money in terms of design, proposed timetable and personnel issues. It took a year before financial closure, longer than expected, due to changes in proposed design and contractual issues.

The case also indicated the need for standardization. Despite the long history of concession structures in the UK, little has been learned from other projects. The case of the West Middlesex hospital was one of the pilots for the standardization operation and has been the first deal closed with these standards.

5.5 PPP in infrastructure: Europe

Many other cases could have been described to illustrate the widespread areas of application of concession contracts in Europe. Every single case has its own story and history, and the impact on value for money will depend strongly on the situational characteristics. The cases described in this chapter differ in motives, procurement rules, selection criteria and added value, and were chosen because they illustrate the various types of concessions in Europe together with the various areas in which concessions can be applied.

5.5.1 Creating concession PPPs

In all three cases, better value for money was the key motive for the public client to start with concessions. Value for money was translated into economic terms. It was expected that a private consortium could deliver more value for less money than could a public agency. Another important motive was risk transfer to the private sector.

Concessions are a way to alleviate lengthy burdens on public budgets. The public budget constraints in combination with urgent need for infrastructure and to stimulate economic growth have been the central motives for the Spanish government to implement concessions as illustrated by the Metro de Sevilla. The strong demand for public transport and the budget situation in Andalucia forced the government to seek private finance. In contrast, in The Netherlands and the UK the urgent need for infrastructure is not a primary motive (at least officially). Reduction in time and costs as well as generating more value in terms of design and services were the political arguments in the case of the West Middlesex hospital. In the case of the school project in The Netherlands, the motives were a decline of maintenance and exploitation costs in combination with quality improvement, and the fact that the teaching staff could direct their attention to education.

For the public client, economic factors are the key motives for starting concessions. In the case of Montaigne Lyceum other aspects such as innovation and quality improvement were indicated as important impacts of the concession contract, but in the tendering procedure focus was on financial revenues and risk transfer. In the case of West Middlesex hospital, increasing flexibility, besides economic reasons, was an important motive. In all three cases, selection of the private partner took place on economic grounds.

For the private bidders, economic expectations concerning revenues and risks were the key drivers to bid. The sociological factors mentioned in Chapter 2, for instance reputation and ideology of the bidding consortium, did not play an intrinsic part in selection of the partner. However, in the case of Montaigne Lyceum, building expertise on concession contracts was an important driver for private consortia to submit a bid. Bidders stated that the financial benefits did not outweigh the transaction costs, but building up expertise is important for future projects.

5.5.2 Procurement rules and selection criteria

In Spain, the lowest cost and the urgency of the need for infrastructure are reflected in the restricted time-frame of the bidding process. There is no prequalification and the bidding process before financial closure takes up to four months. Furthermore, the selection criteria are clear and simple. Points are allocated to specific technical (maintenance and level of service) and commercial criteria (tariff level, concession period and subsidies).

In the UK and The Netherlands, the tendering procedure can take from 18 months to two years; the bidding costs are therefore high. For instance, for hospitals in the UK, bidding costs may amount to about €4.5–6 million, whereas in Spain they are estimated to be €0.5 million.[4] An important difference between

---

[4] Based on interviews with concession holders, review of international case studies (Dewulf et al., 2004).

the Metro de Sevilla and, for instance, the Montaigne Lyceum is that in the latter case negotiations started with the four short-listed bidding consortia. Pre-qualification based on project finance, experience and references, followed by an invitation to negotiate is the typical procedure occurring in many western countries that have adapted the UK system. The case of the West Middlesex hospital indicates that the NHS in the UK is making an effort to reduce bidding costs and time: during procurement the NHS went directly from three bidders to a preferred bidder, claiming it reduced both costs and time significantly. In Spain there is no concession contract negotiation process: the contract outline remains unchanged during the tendering procedure.

5.5.3 Performance

As indicated, the cases described should be seen as snapshots, not empirical evidence. The findings cannot be generalized to concession contracts overall. However, they do confirm some empirical findings described in Chapter 4 and give the reader more understanding about how different variables influence performance.

Financial performance

Efficiency
There is a large incentive for the concession holder to take measures to increase efficiency in construction of facilities: the lower the costs, the higher the value. For the West Middlesex hospital, the concession scheme led to a reduction in construction time, with the new hospital being constructed in only 18 months. Other important benefits were the price certainty and the allocation of risk to the private sector.

Another important difference between concessions and traditional pro-curement is that the payment depends on performance, as demonstrated by the case of the Montaigne Lyceum. In traditionally procured contracts, however, the government pays the construction costs. The contract requires the building to be available for educational purposes, for which an 'availability fee' is paid by the public authority.

An additional incentive by which concession holders can increase performance is to introduce a bonus system as in the Metro de Sevilla case. Bonuses can be awarded when specific performance criteria are met during the operating or maintenance stages (safety, waiting times, etc.). Such a bonus could be to lengthen the concession period by one year.

Transaction costs
A major difference between the Spanish case and the two others is the bidding time, and consequently the cost. Protracted bidding times and high costs mean

that it is hard to generate profits from the contract. The West Middlesex hospital case is a good example of how the UK government tried to reduce transaction costs by speeding up the tendering process and standardizing procedures.

Revenues and risks

Risk transfer is a key characteristic of these PPPs. The risks associated with all three projects described are high; since the future is extremely uncertain the clients asked for flexibility. The West Middlesex hospital contract, for instance, clearly shows the importance of flexibility in current concession projects.

Content performance

Value for money

In general, the cases described indicate how difficult it is to estimate value for money. A PSC was used to compare public with private delivery for the Montaigne Lyceum and West Middlesex hospital. In both cases, the responsible public agent reacted positively, but as mentioned in Chapter 4 use of the PSC is criticized by many.

It is clear, however, that the Seville project would not have been constructed (at least in the current situation) without private investment; the concession contract enabled the municipality and the region to start this complex project. This is certainly not the case for the Montaigne Lyceum, where the introduction of the concession contract can be seen as an experiment or pilot by which to learn and not as a necessity. It is hard to draw conclusions regarding the West Middlesex hospital.

Two of the cases show a clear customer focus. In Seville, the introduction of a concession PPP stimulated the private parties to become more customer-oriented. The design and operating system were clearly aligned with the needs of metro users, it being argued that a more user-friendly (safer, cleaner, etc.) metro station would attract more passengers. Unlike many other concession schemes, the end user of the West Middlesex hospital was heavily involved. According to the NHS (2002), medical personnel were involved in the design from the start, so the needs of staff and patients would not be dictated solely by architects, accountants and surveyors. At every level, clinicians had their say on the key components for each department, how they should work and what accommodation was needed, from the number of rooms to the number of beds. For the Montaigne Lyceum it is too early to evaluate what the impact will be for the teaching staff.

Innovation

Innovation has been confined to technical and operational facets. Financial performance (price, risks) is the prime selection criterion, followed by technical performance. The cases described do not indicate that concession contracts have led to innovation in design and engineering.

Life cycle

Concession holders are forced to think in the long term, which is generally seen as a tremendous competitive advantage of concessions over traditional contracts. The cases described are not yet in the maintenance phase and therefore hard to evaluate. However, because the fees the client pays depend on the availability of the facilities, concession holders have been forced to estimate the life cycle value of their investments.

Process performance: public interest

The cases indicate that safeguarding the public interest is primarily the responsibility of the public actor.

Conditions for success

The success of concessions in the cases described depends not only on the approach, procurement rules and selection criteria, but also strongly on economic, cultural and institutional conditions.

The impact of contextual changes is clear. Legislation and cultural–historical tradition determine the way the project is procured, with differences between Anglo-Saxon and Spanish tradition being apparent. Another major contextual influence on the way the concession is procured is market relations. According to Bongenaar (2001) one of the major fallacies concerning governance structures for PPPs is the lack of competitive environment both in the initial stages and during the execution of the project. For the Montaigne Lyceum, 11 potential consortia submitted bids, but in many cases in the UK only a few bidders signed up. This was also the reason why the NHS simplified the bidding process in order to lower the bidding costs, as was the case in the West Middlesex hospital.

It was difficult to show how organizational variables play a role in the creation and functioning of PPPs. The high number of bidders for the Montaigne Lyceum contract resulted from private parties expecting that the government would launch more concession projects in the near future; they wanted to build up their necessary skills and capabilities. The cases in Spain and the UK demonstrate how experienced both the private and public sectors are with concessions.

5.6 Summary

The cases described illustrate the way concession PPPs are created and procured in three different European countries, each with its own traditions and culture. We now have to wait two years or more to be able to evaluate their performance, but even after such a period of time, it will be difficult to draw valid conclusions: we simply cannot predict what would have happened if the facility had been procured in a traditional way.

The cases nevertheless illustrate the rationale behind concessions and also show some new developments in concessions together with ways of coping with the bottlenecks described in Chapter 4, such as standardization to reduce transaction costs, increase of transparency, shortening the bidding process, and other developments. These changes will be further described and analysed in Chapter 8.

References

Andersson, Elffers Felix (2005) *Quick Scan Montaigneproject: Eindrapportage.* Utrecht: Andersson Elffers Felix.

Association of Chartered Cerified Accountants (ACCA) (2004) *Evaluating The Operation of PFI Roads and Hospitals.* Research report no. 84. London: Certified Accountants Educational Trust.

Audit Commission (2003) *Concession in Schools: National Report.* London: Audit Commission.

Bongenaar, A. (2001) *Corporate Governance and Public Private Partnership: The Case of Japan.* Utrecht: Nederlandse Geografische Studies.

Commission of the European Communities (2004) *Green Paper on Public–Private Partnerships and Community Law on Public Contracts and Concessions.* Brussels: European Union.

Dewulf, G., Bult-Spiering, M. & Blanken, A. (2004) *Opportunities for PFI in The Netherlands.* Enschede: P3BI.

Edwards, P. & Shaoul, J. (2003) Controlling the concession process in schools: a case study of the Pimlico project. *Policy and Politics,* **31**, 371–385.

EMTA News (2003) *Quarterly Letter of the European Metropolitan Transport Authorities,* **12**, 3.

Greve, C. (2003) Public–private partnerships in Scandinavia. *International Public Management Review,* **4** (2), 59–68.

National Audit Office (2002) *The Concession Contract for the Redevelopment of West Middlesex University Hospital.* Report by the Comptroller and Auditor General. London: National Audit Office.

Webliography

European Union www.europa.eu.int (accessed June 2005)

NHS Magazine on-line www.nhs.uk/nhsmagazine/archive/feb03 (accessed June 2004)

PPP Knowledge Centre www.pps.minfin.nl/ (accessed June 2005)

6 Joint Ventures
Co-author: Marnix Smit

In urban governance, joint ventures are the dominant type of public–private partnership (PPP). Co-ventures and joint developments are closely linked features of public–private co-operations in this field. Only recently has the joint-venture PPP type of contract been used in other sectors such as healthcare (see Chapter 8). Examples of joint venture PPPs can be found in the USA and in Europe, with many examples in The Netherlands.

Joint-venture PPPs can have project-based and/or policy-based characteristics (see Chapter 2). With joint venture PPPs, the USA and UK show examples of both project- and policy-based partnerships; in The Netherlands and in some other European countries project-based partnerships are used for urban and area development projects.

The London Docklands project is an example of an urban development corporation (UDC) (Bennett & Krebs, 1991). Since 1981 such UDCs have been one of the policy instruments of the Department of the Environment (DoE) for realizing economic development. A UDC is a central government body that tries to achieve economic development in specific areas such as the London Docklands, Manchester, Southampton and Wolverhampton. The budget of a UDC, for example, can be spent on renovation, infrastructure and buildings. For a private party the advantages of a UDC are the speed and security with regard to economic activities and obtaining political support. The advantages for local governments are the additional funding and the concentration of funding in a certain area. Government and market parties need each other to reach their goals. A further example of restructuring in the UK is the so-called English Partnerships (EP), another central government body, 80–90% subsidized and organized regionally. EP is autonomous: it decides on projects within the central government policy basis. However, central government remains in financial control. EP is aimed at urban and 'city challenge' areas, and its goal is to promote economic growth and employment. The focus is on redevelopment of areas (companies, houses, facilities). The most important activities of EP are co-operation between public and private parties, putting investments in place, acquiring foreign companies, and functioning as a knowledge centre.

More examples of joint-venture PPPs with policy-based characteristics will be described in Chapter 7.

In the USA, post-war PPPs in urban governance can also be subdivided into redevelopment partnerships (project based) and growth partnerships (more policy based). Redevelopment partnerships dominated in the 1950s and 1960s as a response to the disinvestment that had undermined the downtowns of many industrial cities, for example the Charles Center in Baltimore (see Chapter 7). In these formal PPPs, with general private partners and local governments as limited partners, responsibilities were clear: both sides benefited, and the effects and revenues of the partnership and the projects were obvious for both the public and private parties. The successors to redevelopment partnerships, growth partnerships, included a broader range of growth sustaining activities in the late 1970s and early 1980s. Specific projects were substituted for larger-scale city renewal, and issues such as long term planning, housing, job training and school reform were also addressed (Beauregard, 1997). With this movement, the shift towards more policy-based PPPs became a fact. In these PPPs, local business leaders or local economic elites were the major private sector participants, with local ties to particular places.

From a theoretical perspective, a dominant approach for studying PPP in urban development in the USA has been urban regime theory; this theory is attractive in understanding PPPs because it makes distinctions based on co-ordination processes. Urban regime theory (Rubin & Stankiewicz, 2001):

- Focuses on the relationship among the public, private and non-profit community sectors in urban politics
- Identifies the private actor as powerful in urban politics.

Urban regimes can be organic, instrumental and symbolic. The instrumental type of urban regime is particularly applicable to PPP; such regimes set short term goals related to specific projects and give priority to tangible results (Stoker, 1997).

However, not all PPPs are regimes: the 1980s examples of governance regimes in American cities have become rarer and over the past twenty years, PPPs that involve special purpose *projects* have gained ground (Austin & McCaffrey, 2002). Because of the growing absence of local business elites, and therefore of private actors with local ties, real estate and construction companies have become the dominant private sector parties in urban development projects. Project-based PPPs are not regimes themselves, although regime theory can help us to understand them. Nevertheless, the dynamics of the formation and operation of PPPs require additional theories to explain them.

In this book, PPPs are regarded as formal arrangements between public and private parties, concerning project-specific efforts to mobilize resources and share risks in order to deliver a certain product; therefore, this chapter focuses on project-based joint-venture PPPs, interesting examples of which are found in complex urban area projects in The Netherlands. The Dutch concept of joint venture originated in the 1980s in inner-city projects. Since then many successful joint ventures between local government and private parties have been launched at a local level. In contrast to the success at local level, far fewer project-based joint venture PPPs involve central government. In recent years, several policy documents have been published regarding the opportunities for joint ventures in complex urban area development.

A good example of a local joint venture PPP is Nijmegen–Marienburg. This project started as part of a plan for revitalizing the city centre of Nijmegen. The project consisted of integrated development functions such as shopping, living, working, culture, services and parking. A partnership between the city of Nijmegen and the real estate development division of the financial services company ING developed an entirely new area for shopping and living in the centre of the historic city of Nijmegen. The development of the area resulted in:

- 74 houses
- 12 500 m^2 of office space
- 4500 m^2 for a library
- 4500 m^2 for the archives
- 3500 m^2 for a cinema club and theatre
- 14 500 m^2 of shops and catering facilities
- 15 000 m^2 of public open space.

The magnitude and the accompanying risks made the city council of Nijmegen decide to develop the project in co-operation with an experienced private actor. The project is generally considered a very successful example of PPP and is noted as an example of intensive use of space within the framework of sustainable urban development (Bult-Spiering, 2003).

Another example of a successful local joint-venture PPP is the revitalization of the city centre of Den Bosch. Bordered by the historic city centre, an urban renewal project has been realized. This project was initiated by constructor/developer Royal Volker Wessels Stevin and developed in co-operation with the local authority of the city. This joint venture commenced in 1990, the goal of the partnership being to 'Change the area into a high grade urban area that is characterized by a high density with high quality, a mix of different functions and good accessibility.' (Dewulf & Spiering, 2001). The project plan embraced the following elements:

- 170 000 m^2 of office space
- 3500 m^2 of shopping space
- 1500 apartments
- Garaging for 1000 cars.

The joint-venture type of PPP is also used in other sectors, such as waste water management and in the delivery of urban environmental services in developing countries. As we are focusing on the construction sector, those interesting examples are beyond the scope of this book.

6.1 Project-based joint ventures

In joint-venture PPPs, the government(s) and one or several private parties co-operate in the development, and in some cases the maintenance and/or exploitation of the project. The joint venture can take the form of a contractually agreed collaboration, and a joint legal entity with principal authority and with both public and private shares is created. Various different (combinations of) legal structures, such as partnerships, limited partnerships, private limited companies or public limited companies, can be applied. These different structures have their own characteristics with regard to project direction, financial and fiscal aspects, liability, etc. The pros and cons of the structures on these elements and therefore the use of different types of structures will depend to a great extent upon the project characteristics, the context of the project and the legal framework and regulations of the country in which they are applied. A common characteristic for all these forms of collaboration is that parties bring in labour and capital to achieve a common goal. Project direction and risks, revenues and losses are key elements in joint ventures. If no special agreements are made, these elements are proportional to the possession of shares or the contribution of means in the joint venture. This explains the difference from concessions. In joint ventures, project direction, risks, revenues and losses are 'shared', while in concessions these are split up between parties. In joint ventures, the emphasis is on 'togetherness'. (Bult-Spiering & Dewulf, 2002).

6.1.1 Characteristics

In practice, a great number of different joint-venture models are applied. The diversity appears, for example, in the different legal structures, the participating parties and the scope of the project which concerns not only the building programme, but also refers to the question of which project phases the public–private co-operation encompasses. In a number of projects the exploitation of land might be a part of the public–private joint venture, whereas in other projects a municipality might take care of this, and realization and exploitation of objects are the central issues of the joint venture.

Under a joint venture, both the public and private sector partners have invested in the company and therefore both have a strong interest in the success of the venture. Full responsibility for investments and operations gives the public and private sector partners a large incentive to make efficient investment decisions and to develop innovative technological solutions. Early participation by the public and private sector partners allows for greater innovation and flexibility in project planning and helps ensure that both partners are able to optimize their goals. Early dialogue between the public and private sector partners can help reduce the transaction costs associated with more traditional tendering processes (Organization for Economic Co-operation and Development, 2000).

However, an important condition for co-operation in joint ventures is that the goals of the partners do not conflict. Goals of partners have to correspond to such an extent that the partnership's interest can be looked after jointly. From this point of view, a sub-condition for the project is that it generates money. The returns of a PPP in complex urban area development are different for each participating actor and consist of different elements. These elements can be expressed in four kinds of added value (see Section 2.2.1): added value in content, added value in process, financial added value and external added value.

6.1.2 Motives

In joint ventures, the focus is on sharing risks and revenues. In the terms of Bennett et al. (2000) (see Chapter 1) the emphasis is on co-responsibility and co-ownership. The motives for both public and private partners to enter a joint venture are then different from concessions where the focus is on transferring risks and revenues. The co-operation between the partners in a joint venture takes place in a separate legal entity. Several arguments can be distinguished for this type of co-operation (Bregman, 1990):

- A separate legal entity for the implementation of the project, so parties do not have separate decision-making processes for legal action with regard to the PPP parties. For exploitation, a separate legal entity is most practical
- A recognizable organization, so all parties have an point to turn to whatever way they are involved in the project
- Early adjustment of goals and interests
- Public and private project direction
- Public and private risk *sharing*
- Partners have limited liability, so they can only be held liable to a certain extent and not for their own separate property
- It appears that separate legal entities are more likely to receive subsidies than bodies formed by contractual co-operation agreements
- A joint venture might offer fiscal advantages. This is, for example, the case in a legal structure that is often used in Dutch urban development projects in which a private limited company (BV) having both public and private shares, controls a limited partnership (CV). This offers fiscal transparency. Parties remain under their own tax regimes and, as a result, profits or losses are passed on to the participating parties. Governments do not pay taxes and private partners can balance potential losses at the start of the project with company profits.

6.2 Complex urban area development

Joint ventures are primarily applied in urban projects, especially in inner-city developments. To be able to analyse the characteristics of the creation and functioning of these joint ventures we have to understand the specific characteristics of urban projects. Urban area (re)development projects are designed to solve problems about liveability and accessibility. Most projects are of local or regional importance. The users of the new facilities are quite easy to identify. The key roles of the private actors are development or exploitation, which explains their involvement from the early stages of the projects throughout exploitation and management. Private partner selection is based on project vision, plans, feasibility studies and experience. Co-operation is embedded in declarations of intent and co-operation agreements. The project organization comprises a steering committee, a project team and several study groups; other stakeholders are merely involved in external participation groups.

In complex urban area projects, where different functions have to be combined, joint-venture PPPs are regarded as a way to generate optimal use and integration of different functions (see, for example, Ernst & Young Consulting, 2000; Centraal Planbureau, 2001; Ministry of Finance, 2002). These complex urban (re)development projects have a number of characteristics: some of these features refer to the content of the project and others refer to the development process.

- Area development instead of object development
- Multifunction approach: integration of different functions, such as housing, infrastructure, retail, office space and open space
- Large-scale financial investments
- Long and dynamic planning and development process
- Multi-actor approach: local government, developers, neighbours and owners are the main actors involved
- Complexity: both complexity of content, and horizontal organizational complexity, because of the multi-function and multi-actor approach
- Added value created through the higher quality of the solution, adjusted goals and interests, the synchronization with developments in adjacent areas, and the dispersal of risks.

Scarcity of space, due to an increase in the demand for different functions, implies that spatial and economic infrastructure needs large investments (Ministry of Economic Affairs, 1999; Ministry of Housing, Spatial Planning and Environment, 2000). The challenge is to combine the quantitative demand with an increase in quality. Urban development strategies that ought to meet these ambitions are the following (Ministry of Housing, Spatial Planning and Environment, 2000):

- Intensifying the use of space
- Combining functions
- Transformation of space.

These ambitions require adequate interaction between the different parties involved in spatial development. The Dutch government has indicated that an integrated approach towards areas, in co-operation with involved public and private parties, offers opportunities to increase quality and speed up development organizations. For example, the Dutch Advisory Council for Transport, Public Works and Water Management (1998) and the Dutch Knowledge Centre on PPP (2002) plead for an integrated approach to area development and intensive co-operation between public and private sector parties.

Therefore, complex urban area development requires both a *multi-function* and a *multi-actor* approach. Public and private actors must work together on specific projects because of their interdependencies. These interdependencies are related to the different functions that need to be established in certain areas and therefore require an integrated approach. The multi-function and multi-actor approaches must be integrated, a need that is met in PPPs.

6.2.1 Multi-function approach

Due to rising population and prosperity in most western countries, the need for space for housing, employment, water, nature and recreation is increasing. Mobility grows as a result of the increasing prosperity, and so does the quantitative demand for roads, railways, waterways, etc. Next to these quantitative space requirements, the qualitative demand for space is also rising. Spatial quality is often defined as diversity, an important determinant of which is a mix of different functions. High spatial quality can be created if these different functions show strong coherence (Sociaal Cultureel Planbureau, 1999).

Today, in most urban projects, various functions are integrated. Growing dynamics and complexity in combination with the lack of space mean that co-ordination and integration of spatial functions will become the challenge for the future. As mentioned in the introduction to this chapter, joint ventures are especially appropriate for complex urban projects.

Complex urban area development is characterized by the integration of various types of function. The Minister of Finance has made a distinction between urban areas, rural areas and combination projects, as shown in Table 6.1. Complex urban area developments need to create and exploit coherences between the different spatial functions. Table 6.2 lists the categories of object that fulfil different possible functions.

Integrated area development is intrinsically complex and requires another role of the public sector. In a study undertaken by P3BI (Blanken et al., 2004), the characteristics of the development of single real estate objects was compared with complex urban area development. Single development focuses on one function, whereas complex urban area development integrates different spatial functions (multi-function approach). Table 6.3 presents the main differences between these two types of development.

Table 6.1   Types of area development (Ministry of Finance, 2002).

Type of area development	Examples
Urban area	New living/working areas Restructuring living areas Waterside projects City centre projects Protected city and village sites New industrial parks Restructuring of industrial parks
Rural area	Reconstruction areas 'Red for green' projects Nature projects Water projects
Combination projects	Coherent development of national infrastructure and the location concerned, e.g. station areas

Table 6.2   Spatial functions and objects (Bult-Spiering et al., 2005).

Main category	Sub-category	
Real estate (commercial and residential)	Houses Manufacturing space Shops and leisure Hotels Schools Hospitals	
Infrastructure	Primary (road, rail, water, air)	National infrastructure Regional infrastructure Local infrastructure
	Secondary	ICT facilities Energy supplies Water supplies (drinking water/waste water)
Public space	Parking Squares Green areas (Recreational) water areas Culture	

All forms of complex urban area development strive for coherent (re)development of physical–spatial elements that are necessary for performing desired functions in a certain defined area. The concept complex is outlined as follows (Bruil et al., 2004):

- The mutual adjustment between different spatial levels of scale
- The mutual adjustment between different policy sectors and experts (including PPP)

- The adjustment between different phases in the process
- The adjustment of physical and spatial elements on technical, judicial, political, economic, ecologic and sociological–cultural conditions.

The Dutch Knowledge Centre on PPP (2003) defines integrated area development as:

'The coordination of different relevant functions (such as living, working, recreation, mobility) and interests (public and private), ending in an overall solution for the area of concern. The best overall solution is the solution that creates an ideal relation between added value and required contributions for all stakeholders.'

Integrated area development is strongly favoured and considered necessary (Dutch Advisory Council for Transport, Public Works and Water Management, 1998; Knowledge Centre on PPP, 2002). Sustainable built environments require an integrated approach because of the interdependencies between the different spatial functions. These interdependencies concern the spatial and functional integration of the project elements, and also the interaction of the project as a whole with its surroundings. Spatial quality seems to connect positively with a high degree of diversification, and an important determinant of diversification is the mixture of different functions (Sociaal Cultureel Planbureau, 1999). More-over, the value of the various parts of the project is very much determined by its relation with its surroundings (Overleg over Ruimtelijke Investeringen, 1991). An integrated approach to urban governance thus tries to create and exploit the coherence between the different spatial functions and objects, assuming that this will lead to spatial, functional and financial synergy.

Besides these strong arguments for an integrated approach, there are of course arguments for separate development of functions. There is a natural tendency to

Table 6.3 Differences between single real estate object and complex urban area development (Bult-Spiering et al., 2005).

	Complexity	Investments	Establishment time	Public sector involvement
Single real estate object development	Simplicity of project content and organization	Single party	Short	Only involved through legislation and rules
Complex urban area development	High complexity of project content and organization	Too high for one single party	Long	Comprehensive co-operation between public and private stakeholders

split up a project into different parts. As the length and the number of functions increases, this tendency increases as well. Splitting up the project is a way of making financial risks controllable and orderly (Daamen, 2005). Transparency is greatest if functions are clearly separated, each with its own products and costs. Performance indicators remain easily traceable. Furthermore, integration of functions leads to less tension in the marketplace. This has to do with two factors: in the first place the number of possible providers will become smaller, and secondly the risk arises that the extra complexity implies that the executive actor will become indispensable (OC&C Strategy Consultants, 2002). Finally, the government is not organized in a way that corresponds best with an integrated approach. Area development projects often contain parts where the political administrative authority resides at different levels and sectors of the government.

The arguments for and against integration of functions imply that optimal co-ordination of spatial functions demands comprehensive co-operation between the various public and private parties involved. The (re)development process must be organized so that synergy opportunities are used most advantageously. There is, however, obvious tension between the striving for coherence on the one hand, and the fragmentation of actors, goals, means, investment opportunities and time schemes on the other. The involvement of many stakeholders, with various interests and goals, causes a high degree of complexity. Therefore, it is necessary to understand both the functions and the actors involved in complex urban area developments.

6.2.2 Multi-actor approach

The co-ordination and integration of multiple spatial functions implies co-operation between the many actors that are involved. Urban area development projects, where joint ventures between public and private parties are used, are complex open-ended projects which makes them hard to manage. Most building processes are open systems in which different actors are involved. Planning involves searching for common goals and solutions, and settlements with regard to costs, risks and control mechanisms, for example. The development process in these projects is characterized by a high degree of uncertainty and ambiguity with regard to the product and the way it is realized. Various public parties are involved together with different private parties such as land or real estate owners, landlords, retailers, users, developers and other private persons. The greater number of parties carrying responsibility for parts of a project is an important difference between complex urban area development and object development. The fragmentation of means, such as land, capital, real estate and authority, causes the various parties involved in the area development to be mutually dependent. No party is capable of realizing development of the area by themselves, but all provide a part of the puzzle. This means that there is a great number of divergent interests and aims, among which there is often not even an unambiguous public interest, but a complex of divergent and possibly conflicting public interests.

Attaining the quantitative and qualitative ambitions in spatial development by means of an integrated approach towards areas thus requires adequate interaction and co-operation between the different parties involved.

Public sector

If we take a closer look at the public actors involved in area development projects, a distinction can be made between:

- Projects in which both the central government and local or regional government have authority over certain parts of the project (multi-level)
- Projects in which only the local government has direct authority.

In the first type, the area development project consists of various parts with the political administrative authority distributed among different levels and sectors of government. Various central government departments, such as the Ministry of Transport, Public Works and Water Management, the Ministry of Housing, Spatial Planning and the Environment, and the Ministry of Economic Affairs, or subdivisions of these departments, might be involved. Examples of large-scale area development projects in which these departments are involved are the so-called new key projects for the new European high-speed rail network (HSL). The central government bodies can fulfil a number of different roles in these developments. Besides a minimum role consequent upon the legal framework, the central government bodies fulfil the role of facilitator in almost every project in which they are involved; carrying out this role also depends on existing regulations with regard to subsidies and land ownership. Whether the central government will take on a more active role, such as director or participant, depends on a number of criteria (Dutch Knowledge Centre on PPP, 2003):

- The magnitude of the national interest
- The extent to which this interest can be realized
- The magnitude of the risks
- The ownership of real estate or land
- The physical or direct geographical connection between central and local government interests
- The financial interest of central government.

Traditionally, central government does not usually participate in joint ventures on a risk sharing basis. In most cases, the central government prefers a more distant role and closes an agreement with the local government in which financial contributions and conditions are settled, amongst other things.

Because of the administrative authority of municipalities with regard to spatial developments on their territory, they are the prime participants in area development. This applies to both multi-level area development projects and local area development projects. Most projects are initiated by the municipality, which also makes and ratifies zoning and urban plans. By means of these plans and the land-use policy, the municipality tries to regulate the use of space in a way that creates the best conditions for their citizens: it thus falls to the municipality to reconcile the various different interests such as local transport, housing, employment, culture and recreation. The municipality often takes on the role of director in development processes, (partially) owns the land, and has the ability to formulate conditions for the development of land. Because of this, the municipality is involved from the first stage of the development process and often is contract partner or joint venture partner.

To use the coherence between functions, co-operation between public and private parties is necessary. Actors are mutually dependent, but will (although they have agreed upon a common goal) strive for their own goals and look after their own interests. Often, merely in the execution of plans, the different interests of central government departments become clear; instead of a joint effort the different departments take care of their own interests and different parts of the area are developed with diverging priorities, phasing and use of means. Within central government there is no integral director or 'owner' of the problem. Public actors are often both a partner in the joint venture and a buyer of the product: these double roles might cause a conflict of interest if the public and private, e.g. joint company, interests diverge.

Private sector

Just as there are multiple levels and sectors involved on the public side to look after various, sometimes divergent public interests, the private sector is also differentiated. An important distinction can be made between private parties which participate in project establishment and parties which are involved otherwise. In joint-venture PPPs for complex urban developments the following parties can be involved:

- Project developers
- Contractors
- Investors
- Real estate managers
- Financiers
- Engineers
- Housing corporations.

These different private parties have specific interests in diverse phases of the project. They all want to be involved in the project as early as possible, however, to serve these interests best.

The relationship between actors is characterized by both mutual dependency and dynamics. Dynamics make the behaviour and activities of actors in a network hard to predict. This causes extra complexity for a successful multi-actor approach. The creation and exploitation of interdependencies between functions and actors affects the organizational structures of joint-venture PPPs; these effects are described in Section 6.3.

6.3 Organizational aspects of joint-venture PPPs

In Chapter 3, a general outline was given of the way PPPs are procured in Europe and in the USA. In this section, we confine ourselves to some typical organizational characteristics of joint-venture PPPs.

6.3.1 Scope and balance

In joint venture PPPs, the focus is on co-operation. Hence, one important aspect is the way revenues and expenditures are balanced between public and private partners. The revenues will depend on the scope of the project; this is partly determined by the size of the area to be developed. For the limitation of the content of the project, the area involved in the attribution of costs and benefits can be looked at. Normally this limitation will be determined during the development of plans and is often subject to change.

In joint projects between the government and private parties, the government usually grants subsidies to non-commercial functions of the project, while the financial benefits arise from the commercial functions of the project. Because the project uses an integrated approach, balancing between these parts is often applied and is an acknowledgement that the financial benefits of commercial functions also depend on the subsidy granted for non-commercial functions; for this reason profitable and non-profitable parts are linked in complex urban area development projects. For example, the returns on real estate development can be used to finance partially a desired infrastructure solution, because the realization of infrastructure contributes to the value of the real estate. This is also called 'value capturing'. Returns are not withdrawn from the project, but are entirely or partially re-invested, which contributes to a higher quality of plan and makes non-profitable elements of the project realizable.

6.3.2 Phasing and procedures

The way joint ventures in urban projects are organized has to be aligned with the phasing and procedures in urban development. Spatial projects often take a long period from start until final realization. To a great extent, this is because these projects have a large political and societal interest. Frequently, the need for the development is subject to repeated discussion. In complex urban development, infrastructure and urban developments are realized in mutual coherence, implying that the phasing of the infrastructure development and the urban development must be interwoven. For this reason, the phasing of Fig. 6.1 can be taken as the point of departure for complex urban area development projects.

The formal decision-making process in complex urban area development is dominated by the legal procedures through which the project has to go before realization. If central government bodies carry authority for certain parts of the project, for example a large-scale infrastructure development, procedures at both national level and local level have to be executed.

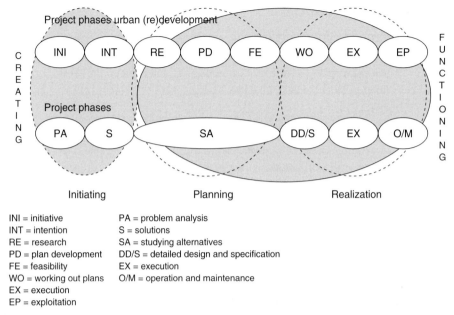

Figure 6.1 = initiative
PA = problem analysis
INT = intention
S = solutions
RE = research
SA = studying alternatives
PD = plan development
DD/S = detailed design and specification
FE = feasibility
EX = execution
WO = working out plans
O/M = operation and maintenance
EX = execution
EP = exploitation

Figure 6.1   Project phasing for a complex urban area development.

In general, the granting of permits for project realization has to take place during a stage in which government and private parties have already closed a contractual agreement on realization. Keeping up the agreement is impossible without granted permits. Often reservations are made in the agreement with regard to the realization of the required legal public basis for the project. The public partner is obliged to make the best effort to get this done.

The legal framework for spatial developments influences joint-venture PPPs. Practice shows these spatial development procedures can be a bottleneck in PPPs. However, the long spatial procedures are not a specific PPP problem. In projects that are realized entirely by the government, spatial procedures also fulfil an important and time-consuming role. Experiences to date show that the success of a PPP does not depend directly on spatial procedures. Nevertheless, PPPs benefit from speeding up these procedures. From the public side, attention is repeatedly paid to these problems by looking for possibilities to speed up procedures and simultaneously safeguard their quality without affecting the democratic principles of administrative responsibility.

Co-operation between partners can concern all phases of the project, or only some of the phases. For each project phase what knowledge and contribution are required and what organization fits best can be determined (see Section 2.2.2).

6.3.3 Procurement and risks

Complex urban area development projects can go through the phase of making plans in different ways. The contribution of private parties can take place in different forms and at different times during project planning. Three possibilities can be distinguished (Bult-Spiering et al., 2005):

■ Government(s) go through the planning phase and draw up a specification for the output, the optimal scope of the project and the desired method of co-operation with private parties, after which procurement takes place
■ By means of a procurement process the government(s) decides at an early stage to co-operate with a private partner: important motives can be
— added value in plans and realization by combination and mutual adjustment of public and private plans at an early stage of the development process
— risks with regard to realization and exploitation are influenced by decisions made in the plan-making phase: it is more efficient to transfer these risks at an early stage to the party that is best capable of controlling them
— the government(s) does not want to realize the plan without a substantial contribution from the private sector; early co-operation when drawing up plans can enhance the commercial characteristics of the project by means of the market knowledge and expertise of private parties – willingness to invest and bear risks increases this way
■ The government(s) believes that, given the specific characteristics of the project, it is desirable or even unavoidable to co-operate with private parties at an early stage; they co-operate for the planning phase and strive for public procurement of separate project parts before realization and exploitation.

In general, government bodies that intend to award a contract for the supply of goods, services or works in excess of a certain value, must do so through the approved process of public procurement as laid down in the EU procurement directives (see Section 3.3.1).

Although selection of a private sector joint venture partner is not subject to public procurement rules, applying these rules might enhance the added value of the PPP. Different ideas regarding the development and exploitation of the project can be assessed and weighted. The public party can, for example, consult the marketplace. To find the best match between needs and possibilities, knowledge of market parties is required; for this reason, the government could consult these parties. However, caution is required. A government that consults extensively with a market party during the planning phase and shows details of the project to this party, faces the risk of having to exclude this party from the procurement procedure. This is the case if that party has an advantage in knowledge and if fair competition between all market parties that are willing to tender is insufficiently safeguarded. Besides fair competition, the protection of their ideas is an important element in consulting market parties. Costs and efforts for potential bidders should be very limited. Based on the consultation, the public party can draw up a general schedule of requirements. This document can serve as a basis for a contest among interested private parties. After having made a selection based on criteria and weights, negotiations with one or more private parties take place which can result in a joint venture PPP agreement.

Contracting

If the public party is the dominant partner in the joint venture PPP, with regard to authority and/or financial input, the joint venture has to be considered a contracting authority. Contracts with third parties, awarded by the joint venture, are subjected to the rules of public procurement. If joint venture PPPs are established on the basis of equal participation by the public and the private sectors, as is the case in many Dutch joint venture PPPs, the crucial element is the extent to which the public party can actually control what happens within the joint venture. If a joint venture PPP that has to be treated as a contracting authority wishes to award a contract to an undertaking which is affiliated with the private sector partner in the joint venture, there is no exception: here it will also be necessary to apply public procurement rules (Dutch Knowledge Centre on PPP, 2001). In short, the contracting authority will have a choice between an open procedure and a restricted procedure or, in special cases, a negotiated procedure (see Section 3.1). Also, if the land to be developed is privately owned, the government body cannot avoid going through a competition procedure (see Section 3.2.2). If the private party is willing to develop the project in a joint venture PPP, and covers all the likely losses with revenues from other parts of the project, the choice of architect, contractor and operator will be on economic grounds and at the discretion of the developer (Dutch Knowledge Centre on PPP, 2001).

Risks

The risks linked to public–private co-operation in urban development can, in general, be divided as follows (Ernst & Young Consulting, 2000):

■ *Administrative and policy risks* These are related to the fact that the public partner gets another governing body during the project which might cause problems regarding spatial procedures, public participation, etc.
■ *Commercial risks* These are related to the supply and demand in the marketplace and estimation of the realization costs and expected returns.

A more specific classification of risks is usually used (see Section 2.2.2).

In joint venture PPPs, both commercial and administrative risks can be estimated and therefore limited. Private parties primarily have knowledge of estimating and controlling risks of commercial real estate exploitation; municipalities have a long track record regarding the production of land and are generally capable of estimating the connected risks.

For almost every PPP project, the public partner is primarily focused on mitigating the administrative and policy risks connected with the project. If the municipality does not cope sufficiently with these risks, not only administrative or policy problems will result, but also commercial risks will be a consequence.

An often-heard idea is that private parties are more capable of coping with risks than the government. This idea primarily concerns commercial risks. Although municipalities do have knowledge in this respect, certainly regarding the production and sale of land, it is acknowledged generally that private real estate developers and contractors are better equipped to make such estimations. This does not mean that private parties are more inclined to face risks or that they are better able to bear these risks. Some municipalities even indicate that their private partners in joint-venture PPPs were more focused on distinguishing, avoiding and controlling risks. Private parties often make agreements with the municipality on how to cope with the extra costs or smaller returns if, for example, the land to be built on is not ready in time (because public participation procedures take more time than expected or the government finds contaminated soil that has to be cleaned first).

In general, the arrangements for handling risks and the distribution of risks are difficult and complex subjects. If risks are not distributed appropriately, significant problems arise; however, sharing knowledge and expertise can mitigate against difficulties. An import potential added value of public–private co-operation resides in sharing knowledge and expertise (added value on process, see Section 2.2.1).

6.4 Performance of joint venture PPPs

Various major bottlenecks arise as a consequence of the complexity of urban area development. A clear tension exists between the drive for a comprehensive approach and the fragmentation between a large number of public and private parties with different interests, resources, perceptions and working methods. The fragmentation of means causes interdependencies between parties: no party is capable of realizing the development by themselves; all should provide a part of the puzzle. This means a great number of divergent interests and aims need to be managed, among which are ambiguous public interests. The difficulties in such management are illustrated by the Utrecht Centrum Project (UCP).

Urban area development projects have a dynamic planning process in which:

(1) At first, the interests of public parties are often not fully clear
(2) Public goals are formulated at a very abstract level
(3) Interests of public actors (sectors and levels) are potentially conflicting
(4) Goals might change due to altered circumstances (technological, economical, political)
(5) A common goal does not exist in the first stages of the planning process.

The Utrecht Centrum Project was aimed at upgrading the station area of the city of Utrecht. The goals were to improve public transport, to improve public space (squares, infrastructure, greenery), to increase multi-functionality, to create a good east–west connection and to use the economic potency of the area.

In 1993, the City of Utrecht, the Dutch Railroad Company, ABP (pension fund and owner of the shopping complex) and Jaarbeurs (private owner of a real estate complex for annual fairs) agreed upon the Masterplan Utrecht City Project that was confirmed by a city council decision. The development company UCP Inc. was set up in which the City of Utrecht (51%) and three private developers participated and several partial projects such as a music centre, office blocks, apartments, hotels, improvement of a shopping centre, road infrastructure adjustments, railway and station hall adjustments, etc., were planned. In 1996 the development company UCP was liquidated as a result of dominance of the city and too little consultation with real estate owners.

Subsequently, in 1996, the administrative platform UCP was set up, in which the City of Utrecht, Jaarbeurs, WBN (formerly part of ABP, owner of the shopping complex) and the real estate division of the Dutch Railroad Company participated. The result was the Final Urban Design UCP that was conditionally (regarding finance and accessibility) authorized by the city council in 1997. The Administrative Platform UCP crashed in March 2000 due to non-transparent partial exploitation of the public facilities, divergent interests in payment, indistinctness about central government's contribution, competitive interests of the retail owners, unsustained political commitment and unclear processes.

The UCP continued in June 2000 with only the City of Utrecht and the real estate division of the Dutch Railroad Company as intended partners. After the city council elections at the end of 2000, this intended partnership was wound up. With its own project organization, the City of Utrecht took the lead in a much smaller project that was now called Station Area. A referendum on two variants of a masterplan took place in 2002. In 2004, central government and the city council both agreed on a new masterplan for the station area. Central government and the City of Utrecht reached agreement for the project, involving a financial contribution of €300 million from central government. Furthermore, declarations of intent for the development of the area were signed by the City of Utrecht, Corio (formerly WBN, owner of the shopping complex), Jaarbeurs and the Dutch Railroad Company. Partners expected to close final development agreements in 2005.

The coherent development of different functions is expected to generate added value. In joint venture PPP practice, public parties consider added value in content and in the process to be most important. Private parties primarily mention financial, process and external added value as being most important. A possible explanation is that exactly these types of added value are linked to the organizational goals of both public and private actors (Bult-Spiering, 2003). Parties do not participate in a PPP only because of an expected added value, but also because of political and public pressure and the general idea that co-operation leads to a better final result (Spiering & Dewulf, 2001).

It is difficult to make valid statements on added value or performance of joint-venture PPPs. The model used is always unique, most joint-venture partnerships are applied at a local level, and performance is often determined by very specific project-based characteristics. This makes comparative evaluative research difficult and means that only a small amount of empirical evidence on the performance of joint venture PPPs is available.

In Section 2.4, we distinguished between process and product performance. Product performance was defined as financial performance and content performance; process performance refers to actors' fit, public interest and behaviour.

6.4.1 Product performance

Government investments are made in functionally separate paths. Often, investments are placed on the basis of functional considerations, with only very limited investments on the basis of opportunities to combine functions; a sectoral approach still predominates. This problem occurs not only during the execution phase, but commences when policy goals are formulated. Each department has its own sector-based policy, means, time schemes and procedures. Whereas all governments strive to act in the public interest, a single focus on their own responsibilities causes a struggle between different government bodies. This might be one of the reasons why successful joint venture PPPs are hard to establish, in particular for projects in which different layers of government have authority over certain parts of the project. The question of what is the best role for the different government bodies involved, and in what way principal authority in co-operation with private sector parties should be given form, is often not answered adequately by the public authorities involved.

As research shows, another source of bottlenecks arises if private parties have to make substantial investments before final private partner selection and revenues can only be acquired after this time (Bult-Spiering & Dewulf, 2002). A PPP that applies to only one phase is possible; however, in all other cases the PPP should concern multiple project phases. The principal–agent problem might apply if parts of the project are outsourced to one of the participating parties (see Contracting in Section 6.3.3).

6.4.2 Process performance

In Chapter 4, we stressed the importance of economic aspects for the development and functioning of concessions. Sociological criteria were of less importance, certainly for the selection of the partner. The primary criteria for selection and for the formulation of the concession contract were economic and/or financial because the relationship between the principal and the agent is not based on co-operation, but rather on making a transaction. In joint venture PPPs, the contract is focused on co-operation. It is therefore not surprising that socio-logical aspects such as trust, transparency and other more intangible aspects are becoming important.

A major bottleneck might arise if the interests of parties participating in the public–private joint venture diverge. Often the process of co-operation starts without clear analysis of the interests of the various actors. An evaluative research of 24 PPP projects by P3BI (Spiering & Dewulf, 2001) shows that indistinctness regarding each other's interests and motives is the most important cause for bottlenecks in the process of co-operation.

In a joint-venture PPP, participants will try to maximize the use of their rights of ownership. However, besides the partnership's interest, the government has to serve the public interest: tension may arise between both interests. So to prevent a conflict of interests later in the process, a minimal condition for co-operation in a joint venture is that the partnership's interest corresponds with the public interest to such an extent that jointly looking after these interests is possible. If a (local) government decides to participate in a joint venture this consideration should be taken into account carefully.

If parts of the project are developed jointly, the individual parties might demonstrate 'free-rider' behaviour, where in order to maximize their benefit, parties will try to minimize their own efforts and profit by the efforts of others. If opportunistic behaviour takes place then the commitment to this party can be a problem.

Finally, an institutional cause for bottlenecks results from the administrative authority for different parts of the project being allocated at different layers and sectors of government (public–public co-operation). A lack of mutual adjustment between different layers and sectors of government can cause bottlenecks for an integrated approach to the area and for the PPP. No legal mechanism exists for mutual adjustment of plans between different layers of government and in prac-tice this leaves a lot to be desired (Centraal Planbureau, 2001; Spiering & Dewulf, 2001; Bult-Spiering & Dewulf, 2002).

6.5 Lessons

This chapter has explained the context of a specific type of PPP, the joint venture, most commonly used in complex urban area development. PPP in these projects is always unique. Even with the same project characteristics, the specific context and location variables will affect the creation and functioning of the PPP. Therefore, an understanding of the variety of actors, functions and their interdependencies in complex urban area development is a necessary prerequisite for the management of these complex processes.

6.5.1 Conditions for success

Based on an extensive analysis of the results of case studies in the Netherlands, Nijkamp et al. (2002) identified a number of critical drivers of successful PPP projects for urban redevelopment:

- A joint venture model gives a PPP arrangement a high chance of becoming successful
- A larger scope (geographical orientation) of a development project tends to increase its performance (for example, through a broader marketing strategy)
- Clear, timely and transparent mapping of all costs, revenues and profitability aspects of a PPP project is a necessary precondition
- Clear insight into the planning of project parts, the risk profiles involved and the ways in which actors are involved in different project parts, is critical for the good performance of an urban development project.

From a more sociological perspective the concepts of commitment, trust, acceptance and respect are important for the functioning of the co-operation. These sociological concepts primarily apply to the personal relationships in the co-operation. Personal 'fit' is considered especially important for the good functioning of the co-operation. Empirical data shows, furthermore, that the internal relationships between public and private organizations are also a decisive factor in determining the extent to which the previously mentioned concepts are present. In particular, the existing internal public relationships can have a negative influence on the presence of the sociological concepts of commitment, trust, acceptance and respect, and also on flexibility and perseverance (Bult-Spiering, 2003).

Different joint-venture models can be used to create added value. Of significance are what actors can mean to each other, and what their motives and complementary skills are. These motives and skills influence the point at which the partnership should start.

Public–public agreements concerning scope and output must be made, at least at a general level, before involving the private sector. However, early private involvement is necessary to create optimal possibilities for added value.

6.5.2 Improvements

The problems mentioned in Section 6.4 are difficult to cope with, because many of them arise from the typical nature of complex urban development projects, with interdependencies between many functions and actors. Some of the pitfalls, however, may be avoided (Dewulf & Spiering, 2001; Evers et al., 2002) by:

(1) Having a clear understanding of each other's interests
(2) Keeping alert for changes in strategies among the various players
(3) Choosing the right organizational structure. Clear and detailed arrangements have to be made at the beginning of the project
(4) Being flexible during the process. As the targets of the different players will change, flexibility is required
(5) Not forgetting the user and paying attention to the targets of the financial parties
(6) Using a multi-function approach when defining a PPP project. Combination and integration within a project improve the possibilities for generating revenues and therefore improve the possibilities for PPP
(7) Thinking in terms of goals, interests, motives and intentions of the many actors involved, instead of in terms of projects and finances.

6.6 Summary

This chapter has concentrated on joint venture PPPs. A general distinction was made between project-based and more policy-based joint venture PPPs. The chapter focused on project-based joint ventures, because in this book PPPs are regarded as formal arrangements between public and private parties, with project-specific public and private investments, in order to deliver a product.

Project-based joint-venture PPPs are primarily used in complex urban (re)development projects. The interdependencies between different spatial functions (multi-function approach) and the fragmentation of means imply that co-operation between involved public and private parties is necessary (multi-actor approach); this co-operation often takes the form of a joint venture. Various elements, such as the pros and cons of joint-venture PPPs, regulations regarding procurement, organizational aspects and, finally, performance issues have been addressed in this chapter. We concluded by drawing a few lessons for joint venture PPPs. It is most important, however, to realize that these PPPs are always unique and have to be tailor-made for each specific project.

Whereas in this chapter the focus was on project-based PPPs, in Chapter 7 two examples of joint venture PPPs with policy-based characteristics will be outlined. Chapter 7 highlights certain aspects of the American urban development context in relation to PPPs.

References

Austin, J. & McCaffrey, A. (2002) Business leadership coalitions and public–private partnerships in American cities: a business perspective on regime theory. *Journal of Urban Affairs*, **24** (1), 35–54.

Beauregard, R.A. (1997) Public–private partnerships as historical chameleons: the case of the United States. In: J. Pierre (Ed) *Partnerships in Urban Governance: European and American Experiences* (pp. 52–70). London: Macmillan.

Bennett, E., James, S. & Grohmann, P. (2000) *Joint Venture Public–Private Partnerships for Urban Environmental Services.* New York: Public Private Partnerships for the Urban Environment (PPPUE).

Bennett, R.J. & Krebs, G. (1991) *Local Economic Development: Public–Private Partnership Initiation in Britain and Germany.* London: Belhaven Press.

Blanken, A., Bult-Spiering, M. & Dewulf, G. (2004) *Publiek–Private Samenwerking bij Integrale Gebiedsontwikkeling.* Enschede: P3BI.

Bregman, A.G. (1990), *Bestuurlijk–Juridische Aspecten van PPS.* Deventer: Kluwer.

Bruil, I., Hobma, F., Peek, G.-J. & Wigmans, G. (2004) *Integrale Gebiedsontwikkeling: het Stationsgebied's Hertogenbosch.* Amsterdam: Uitgeverij SUN.

Bult-Spiering, M. (2003) *Publiek–Private Samenwerking: de Interactie Centraal.* Utrecht: Lemma.

Bult-Spiering, W.D. & Dewulf, G.P.M.R. (2002) *P3BI in PPS: een Visie.* Enschede: P3BI.

Centraal Planbureau (CPB) (2001) *PPS een Uitdagend Huwelijk.* The Hague: Centraal Planbureau.

Daamen, T. (2005) *De Kost Gaat voor de Baat Uit; Markt, Middelen en Ruimtelijke Kwaliteit bij Stedelijke Gebiedsontwikkeling.* Amsterdam: UItgeverij SUN.

Dewulf, G. & Spiering, W.D. (2001) Public–private partnership: the difference between innercity and infrastructure projects. Paper presented at the *7th International Conference on Public–Private Partnerships: The Enterprise Governance,* Enschede.

Dutch Advisory Council for Transport, Public Works and Water Management (1998) *Ambities Bundelen, Advies over de Inpassing van Infrastructuur.* The Hague: Ministry of Transport, Public Works and Water Management.

Dutch Knowledge Centre on PPP (2001) *PPP and Public Procurement.* The Hague: Ministry of Finance.

Dutch Knowledge Centre on PPP (2002) *Progress Report.* The Hague: Ministry of Finance.

Dutch Knowledge Centre on PPP (2003) *Rijksbetrokkenheid bij Integrale Gebiedsontwikkeling en PPS.* The Hague: Ministry of Finance.

Ernst & Young Consulting (2000) *Inventarisatie Faal en Succesfactoren van Locale PPS-Projecten.* Utrecht: Ernst & Young Consulting.

Evers, F., Schaaf, P. van der & Dewulf, G. (2002) *Public Real Estate: Successful Management Strategies.* Delft: DUP Science.

Ministry of Economic Affairs (1999) *Nota Ruimtelijk Economisch beleid: Dynamiek in Netwerken.* The Hague: Ministry of Economic Affairs.

Ministry of Finance (2002) *Het Gebied Verkend: het Rijk en PPS bij Gebiedsontwikkeling.* Tweede Kamer, Vergaderjaar 2002–2003, 28 753–1. The Hague: Tweede Kamer.

Ministry of Housing, Spatial Planning and the Environment (2000) *Fifth National Policy Document on Spatial Planning.* The Hague: Ministry of Housing, Spatial Planning and the Environment.

Nijkamp, P., Burch, M. van der & Vindigni, G. (2002) A comparative institutional evaluation of public–private partnerships in Dutch urban land-use and revitalisation projects. *Urban Studies,* **39** (10), 1865–1880.

OC&C Strategy Consultants (2002) *Samen Werken aan de Weg*. The Hague: OC&C Strategy Consultants.

Organization for Economic Co-operation and Development (OECD) (2000) *Global Trends In Urban Water Supply and Waste Water Financing and Management: Changing Roles for the Public and Private Sectors*. Paris: Organization for Economic Co-operation and Development.

Overleg over Ruimtelijke Investeringen (ORI) (1991), *Publiek–Private Samenwerking bij Grootstedelijke Projecten*. The Hague: Overleg over Ruimtelijke Investeringen.

Rubin, J.S. & Stankiewicz, G.M. (2001) The Los Angeles community development bank: the possible pitfalls of public–private partnerships. *Journal of Urban Affairs*, **23** (2), 133–153.

Sociaal Cultureel Planbureau (1999) *Sociaal Culturele Verkenningen*. The Hague: Sociaal Cultureel Planbureau.

Spiering, W.D. & Dewulf, G.P.M.R. (2001) *Publiek–Private Samenwerking bij Infrastructurele en Stedelijke Projecten*. Enschede: P3BI.

Stoker, G. (1997) Public–private partnerships and urban governance. In: Pierre, J. (Ed) *Partnerships in Urban Governance: European and American Experiences* (pp. 34–51). London: Macmillan.

Further Reading

Bult-Spiering, M., Blanken, A. & Dewulf, G. (2005) *Handboek PPS*. Utrecht: Lemma.

Commission of the European Communities (2004) *Green Paper on Public–Private Partnerships and Community Law on Public Contracts and Concessions*. Brussels: European Union.

Davis, P. (1986) *Public–Private Partnerships: Improving Urban Life*. New York: Academy of Political Science.

Dewulf, G., Bult-Spiering, M. & Blanken, A. (2004) *Opportunities for PFI in The Netherlands*. Enschede: P3BI.

Fosler, R.S. & Berger, R.A. (1982) *Public–Private Partnership in American Cities: Seven Case Studies*. Lexington: Lexington Books.

Moret, A. (2002) PPS en Europees aanbesteden. *Ondernemersbrief*, **4** (www.cmsderks.nl).

7 Examples of Joint Venture Projects in the USA

Co-author: Gerrit-Jan Knaap

The USA has a rich history of applying public–private partnerships (PPPs) to urban planning. In many US cities, PPPs are used as tools for promoting economic development.

President Carter's administration brought PPPs to the forefront in 1978 with the National Urban Policy. Before this, urban problems were increasing in complexity and intensity. Homelessness, soaring property values, urban decay and fiscal crisis were threatening basic public services in cities (Lyall, 1986). It was becoming apparent that the government could no longer solve these problems by itself; the private sector had to be involved (Fosler & Berger, 1982). Private players were seen as being able to provide resources as well as experience and knowledge (Linder, 1999; Rubin & Stankiewicz, 2001). This crisis created new opportunities for the private and non-profit sectors to play equal roles in city redevelopment. This new reality was articulated in the National Urban Policy, which focused on both the social and economic problems of older central cities. This policy document highlighted the role of the public sector in creating or improving market conditions that would entice private investment.

After World War II, as in most developed countries, the population of the USA grew enormously. With increased prosperity, there was also an increase in the amount of space and infrastructure that each person used, and therefore a greater dependency on motor vehicles which brought new levels of congestion and pollution to urban areas, eventually forcing the wealthy to move to suburbs or the countryside. The phenomenon of relocation outside urban centres was eventually termed 'sprawl'. Public awareness of sprawl eventually resulted in the emergence of the field of 'growth management'.

In this chapter, two fields of joint venture PPPs in urban area development are highlighted: older central city redevelopment and transit-oriented development. Both fields have characteristics of project-based partnerships and policy-based partnerships.

7.1 American urban and regional policy

The first wave of growth management took place during the 1960s and 1970s and focused on the environmental impacts of growth. During this wave, aspects of growth initially handled locally were handed over to state and regional authorities (Burchell et al., 2000). By the 1980s and 1990s, growth management had expanded to include issues related to infrastructure, housing, economic development, community character and quality of life (Burchell et al., 2000). Next came 'smart growth', a movement building off the earlier two waves yet more in favour of growth than before (Burchell et al., 2000). This smart growth movement had ties to other dominant planning theories of the 1990s such as 'new urbanism'. To some academics, new urbanism is the 'leitmotiv' of smart growth (Burchell et al., 2000; Knaap, 2002; Janssen-Jansen, 2004).

Growth management, smart growth and new urbanism share similar goals, although they vary in terms of specific applications. In Table 7.1, the main principles of growth management, smart growth and new urbanism are explained, organized around seven main elements: sustainability, organization, existing values and features, public interest and community, future needs, design, and functions.

Table 7.1   Main principles of growth management, smart growth and new urbanism (Knaap 2002; www.cnu.org; www.newurbanism.org).

Element	Growth management	Smart growth	New urbanism
Sustainability	Encourage full utilization of existing facilities	Make efficient and effective use of land resources and existing infrastructure by encouraging development in areas with existing infrastructure or capacity to avoid costly duplication of services or new construction	
Organization	Control the development of new areas to ensure their co-ordination with existing and proposed facilities	Encourage stakeholder participation rather than conflict	The region is the overall context for all planning and this planning should be participative
		Make development decisions predictable, fair and cost-effective	Towns and cities within a region need a comprehensive metropolitan strategy to prosper
Existing values and features	Promote aesthetics and preserve historic and cultural features	Maintain a unique sense of place by respecting local, cultural and natural environmental features	Towns should develop in the appropriate style for their surroundings, while respecting their neighbourhoods

Table 7.1 (*cont'd*)

Element		Growth management	Smart growth	New urbanism
Public interest and community		Promote public safety	Enhance access to equitable public and private resources for everyone	Ensure that everyone in the neighbourhood has access to the necessities of life
		Preserve the character of the community and promote community identity	Promote the safety, liveability and revitalization of existing urban and rural community centres	Make dense neighbourhoods feel nice and open
Future needs		Provide for flexibility to meet future needs	Provide staged and managed growth in urban transition areas with compact development patterns	
Design		Prevent sprawl		Development patterns should not blur or eradicate the edges of the metropolis
				Wilderness, farmland, villages, town edges, town centres, city neighbourhoods and city centres each have their own building densities, street sizes, and appropriate mixtures of retail, residential and other functions
Functions	Housing	Improve housing opportunities, increase diversity and promote better housing developments	Provide a mix of land use to create a mix of housing choices and opportunities	Provide a range of housing choices
	Transport and infrastructure	Reduce traffic congestion and improve road system		Create diverse, safe, comfortable, walkable neighbourhoods
	Farmland and open space	Conserve agricultural land and preserve open space	Conserve open space and farmland, and preserve critical environmental areas	Clear boundaries for towns and cities; preserve land between towns as open space, wilderness or farmland
	Environment	Avoid environmental problems		Minimal environmental impact of development and its operations
	Services	Maintain or improve the level of community services		

7.1.1 Smart growth

Smart growth focuses on land use policies and seeks to concentrate and mix land use types to best support urban development patterns that are efficient and effective.

> 'The redirection of a portion of growth to the inner-metropolitan area, combined with a more controlled movement outward, would consume far less capital and fewer natural resources and enable the achievement of more ambitious development goals. This can be achieved by a concerted effort to rebuild inner-suburban and central-city markets through infill and redevelopment.'   (Burchell et al., 2000)

It addresses three interrelated problems: the density of development, the spatial separation of land use and the lack of transportation mode choice (Knaap, 2002).

Smart growth works to preserve farmland, protect the environment, reduce infrastructure costs, promote human health and contribute to social equity (Knaap 2002). Key characteristics of smart growth include the following (Burchell et al., 2000):

- Control of outward movement/growth
- Inner-area revitalization
- Design innovations
- Land and natural resource preservation
- Transportation reorientation.

7.1.2 New urbanism

New urbanism focuses on the design of the built environment and seeks to reconnect the city, making it sustainable, efficient, equitable and liveable (Newman & Kenworthy, 1996). New urbanism is shaped by the broadly defined principles of walkability, connectivity, mixed use and diversity, design and sustainability.

Walkability

- Most destinations are within a ten minute walk of home and work
- Pedestrian-friendly street design (buildings close to the street; porches, windows and doors; tree-lined streets; on-street parking; hidden car parks; garages in rear lane; narrow, slow-speed streets).

Connectivity

- Interconnected street grid network disperses traffic and eases walking
- A hierarchy of narrow streets, boulevards and alleys.

Mixed use and diversity

- A mix of shops, offices, flats and housing types on site. Mixed use within neighbourhoods, blocks and even buildings
- Buildings, residences, shops and services located in close proximity for ease of walking, to enable a more efficient use of services and resources, and to create a more convenient, enjoyable place to live
- A network of high-quality trains connecting cities, towns and neighbourhoods together
- Attracts a diversity of people – ages, classes, cultures and races.

Design

- Emphasis on beauty, aesthetics, human comfort and creating a sense of place
- Special placement of civic uses and sites within community; human scale architecture and beautiful surroundings nourish the human spirit
- Neighbourhood structured with a discernable centre and edge, public open space at centre, designed as civic art, transect planning where the highest densities are at the town centre, getting progressively less dense towards the edge.

Sustainability

- Minimal environmental impact of development and its operations
- Higher quality of life.

New urbanism brings residents, businesses, developers and municipalities an improved quality of life, increased revenues (through economies of scale), less congestion and sprawl, and greater public satisfaction and safety (www.newurbanism.org).

One of the primary strategies of smart growth and new urbanism is 'transit-oriented development'. Where smart growth and new urbanism are urban development movements, TOD is an instrument that aspires to achieve the goals of these policies in applications related to transportation. TOD is considered by the Capitol Region Council of Governments (2002) to be a smart growth strategy, because it both tackles the issue of where growth should occur from a regional 'sustainability' perspective and co-ordinates land use and transportation, such that both land and infrastructure are used efficiently.

7.2 Joint ventures in city redevelopment: the case of Baltimore

In the 1950s, Baltimore's inner city experienced a decline in population. Like most older industrial cities, neighbourhoods continued to decline and an exodus of white middle-class citizens to the suburbs began. Wartime overcrowding, housing shortages and the decentralization of traditional manufacturing accelerated this trend. The federal Urban Renewal Program provided funds for redevelopment, and the Redevelopment Commission was formed to carry out the programmes in an integrated way. It co-ordinated the activities of the city housing authority, the housing bureau of the city department of health, the bureau of sanitation, the police department, the fire department, and building inspectors in the renewal neighbourhoods. By 1956, the city's capacity increased through the formation of the Baltimore Urban Redevelopment and Housing Agency (BURHA), a formal merger between the housing authority and the Redevelopment Commission. In 1968, the Department of Housing and Community Development (HCD) was formed by merging BURHA with the city bureau of building inspection (Lyall, 1982).

Since 1955, the Greater Baltimore Committee has sought to make the region more competitive by organizing its corporate and civic leadership to improve the business climate. The economic alliance of the Greater Baltimore Committee is the sole PPP that brings together business, government and educational institutions to attract new business and investment to the region (www.gbc.org). The needs of the city, as defined by the Greater Baltimore Committee's 1955 founders, included the accelerated construction of the Jones Falls expressway, the building of a civic centre for conventions, an increase in cultural activities and sports events, the development of the city's port facilities, the development of a plan for modernizing mass transport, and the creation of a comprehensive urban renewal plan (Greater Baltimore Committee, 2005).

The Citizens Planning and Housing Association (CPHA), a single non-profit-making organization, has served as an important catalyst in stimulating institutional change in public agencies and citizen organizations and in the creation of both HCD and the Greater Baltimore Committee (Lyall, 1982). The association is now a member of the Baltimore Regional Partnership, an alliance of five civic and environmental groups, that seeks to ensure that state and federal public funds are channelled to urban redevelopment projects that support transit-oriented development within the principles of Maryland's smart growth legislation (www.balto-region-partners.org/).

The joint ventures of the Greater Baltimore Committee (general partner) and local governments (limited partners) to plan for and invest in the city have resulted in developments in the city centre with impacts on the larger metropolitan area. The Charles Center and Inner Harbor developments were part of redevelopment plans started in 1959 (Lyall, 1982). The Charles Center is a 133 500 m^2 project located in the centre of the commercial district. The site consists of 185 800 m^2 of office buildings, 700 flats, 700 hotel rooms, a theatre, walkways, retail shops and underground car parks. The total cost was US$130 million, including US$20 million for a federal building, US$25 million from a public-bond issue for land acquisition contributed by the city and US$85 million in private development resources.

The Inner Harbor project encompasses the waterfront and the neighbourhoods that surround it. The Inner Harbor plan includes development surrounding the harbour (Harborplace, Maryland Science Museum, National Aquarium, World Trade Center, Convention Center), a campus of the Baltimore Community College, parking facilities and housing. This project cost US$260 million over 30 years, including US$180 million of federal funds, US$58 million in city resources and US$22 million in private investments (Lyall, 1982).

Joint venture

The downtown redevelopment grew out of the master plan which set the priorities and basic principles for the various projects. Baltimore's Charles Center was largely a private sector initiative, whereas the Inner Harbor was primarily a public sector initiative, requiring considerable government finance. The public sector subsidies were justified by the projects' inducements to private sector investment, real estate values and job growth (Lyall, 1982).

The partnership started with the Charles Center and evolved into a sophisticated public–private corporation for the entire Inner Harbor area (Lyall, 1986). The Charles Center-Inner Harbor Management Corporation was formed in 1965 for the purpose of entering into an exclusive contract with the mayor and city council for the planning and development of the Charles Center and Inner Harbor projects; it performed all the activities that a city renewal agency would normally perform in project development and operated independently with a board of directors, but its work and activities were under the direction of the BURHA commissioner and the mayor. This private corporation's sole client was the City of Baltimore; its funding came from the city, its budget was included within the city budget under the Urban Renewal and Housing Commission, and it advised on policy. It did not have power to enter into contracts with builders, developers or designers, but it did have single authority on issues related to those projects. Public and private sector employees, developers, contractors and citizens knew that issues concerning significant urban renewal projects were to be directed to the Charles Center-Inner Harbor Management Corporation. In the early 1990s, Charles Center-Inner Harbor Management Corporation was incorporated into the Baltimore Development Corporation, a new quasi-public non-profit-making corporate entity which was given a broader economic development responsibility and charged with the mission of catalysing the creation of a robust, sustainable economy (www.gbc.org).

Although the working relations between the public and the private partners was not always smooth, the principle of co-operative planning and action was established early, making long term development efforts successful. In the Baltimore example, the personal commitment of individuals was crucial for overcoming the institutional differences between the public and private sectors. The public sector has the capacity to employ procedural and planning tools and political will, whereas the private sector brings in private capital and can act quickly to execute plans. Project complexity means that partnerships require flexibility; PPPs need a stable network of equal interpersonal relationships, stable long term local political leadership and a shared sense of urgency (Lyall, 1982).

7.3 Joint ventures in transit-oriented development: the case of Portland

Travel patterns in the USA are defined by motor vehicle dependency, with three-quarters of the population owning their own cars (Cox & Alm, 2002); this dependency leads to congestion, accidents, sprawl, loss of open space, physical and social isolation, and environmental pollution (Newman & Kenworthy, 1996; Cervero, 1998; Belzer & Autler, 2002a). Although other travel options exist, increasing incomes, pricing policies, declining levels of service and fragmented regulation have made public transport less enticing over time (Cervero, 1998).

Transit-oriented development (TOD) is a strategy within the smart growth and new urbanism movements, designed to address problems created by these travel patterns in the USA. TOD contributes to sustainable regional development in the following ways (US Department of Transportation, 1995; Cervero, 1998; Freilich, 1998; Zykofsky, 1998/1999; Capitol Region Council of Governments, 2002):

- Transport systems are made economically viable
- Public transport is made a more attractive alternative
- More accessible transport facilities are established
- Deteriorating neighbourhoods are revitalized
- A wider variety of housing choices is provided
- Community character is preserved
- Pedestrian activity is increased
- Financial costs of motor vehicle dependency are reduced
- Opportunities for market supported infill (re)development are created.

7.3.1 TOD characteristics

Transit-oriented development (or transit-oriented design) is similar to smart growth, new urbanism and PPP in that it is hard to give a straightforward defini-tion. According to the TOD website (www.transitorienteddevelopment.org),

> 'Transit-Oriented Development is the exciting new growing trend in creating vibrant, liveable communities. It is the creation of compact, walk-able communities centered around high quality train systems. This makes it possible to live a higher quality life without complete dependence on a car for mobility and survival.'

There are some who define TOD by listing its characteristics. Freilich (1998), for example, uses a definition that describes the five main characteristics of TOD:

- Sufficient density to encourage the use of public transport
- Locates residences, jobs and retail destinations close to public transport facilities
- Consists of mixed uses, with retail and employment locations within walking distance of residential areas
- Built on a grid transportation network, which is not divided into the arterial–collector–local road classification system found in most suburban areas
- Contains urban design guidelines and design features to encourage a more pedestrian orientation, which, theoretically, encourages its residents to use public transport rather than their cars.

Offering a slightly different view, Newman and Kenworthy (1999) define the components of TOD as being:

- Walkable design with pedestrians as the highest priority
- Train station as a prominent town centre feature
- A regional node containing a mixture of uses in close proximity, including office, residential, retail and civic uses
- High-density, high-quality development within a ten minute walk circle surrounding the train station
- Collector support transport systems, including trolleybuses, trams, light rail systems and buses
- Designed to include the easy use of bicycles, scooters and rollerblades as daily support transport systems
- Reduced and managed parking inside a ten minute walk circle around the town centre/train station

They list the benefits of TOD as follows:

- Transit investment having double the economic benefit to a city of highway investment
- Enabling a city to use market forces to increase densities near stations, where most services are located, thus creating more efficient sub-centres and minimizing sprawl
- Enabling a city to be more corridor-oriented, making it easier to provide infrastructure
- Enhancing the overall economic efficiency of a city; denser cities with less car use and more transit system use spend a lower proportion of their gross regional product or wealth on passenger transportation.

Cervero (1998) has identified three dimensions (the '3-Ds') of transit-friendly cities and suburbs:

- Density: transit demand rises with the degree of density
- Diversity: mixing of land uses can encourage use of public transport
- Design: good quality urban design is needed, with grid patterns, well suited to walking.

Definitions, such as the three given above, are criticized for lacking prescriptive guidelines for real TOD projects. Belzer and Autler (2002a), amongst others, define TOD in terms of required performance, focusing on desired outcomes such as location efficiency, value recapture trade-offs, liveability, financial return, choice and efficient regional land-use patterns.

> 'TOD must be mixed-use walkable, location-efficient development that balances the need for sufficient density to support convenient transit service with the scale of the adjacent community.' (Belzer & Autler, 2002b)

A related strategy in transportation policy, often confused with TOD, is transportation demand management (TDM). The purposes of TOD and TDM are similar: to reduce motor vehicle use by increasing the use of public transport. However, while TOD focuses on mixed-use economic development of urban areas, TDM aims to make more efficient use of existing transportation resources by shifting demand. (Cervero, 1998; www.todadvocate.com).

7.3.2 TOD key points

Transit-oriented development calls for high-density, mixed land use that is attractive to pedestrians (Cervero, 1998; Capitol Region Council of Governments, 2002). In this context, mixed use refers to residential or office space that supports uses such as retail, restaurants, entertainment, parks and institutions.

To be successful, TOD has to overcome many barriers and prejudices. TOD proposals are often met with concerns about higher costs, depressed property values and altered neighbourhood character, making the implementation of TOD more complicated. To ensure a successful TOD project, it is critical that key stakeholders understand how their own situation meshes with the TOD designation, what the goals of TOD are, and how the goals of TOD can best integrate with the goals of individual players.

Transit-oriented development, like smart growth and new urbanism, requires organizational structures that bring public and private actors together to work on specific projects and to develop shared long term strategies (Ewing, 1997; Cervero, 1998; Burchell et al., 2000; Belzer & Autler, 2002a). Public policies addressing both the demand for and supply of the issue are not enough to meet TOD goals; involving the private sector in this work opens the door to a myriad of new opportunities and impacts, yet with each new player comes new specific interests, some of which are documented in Table 7.2.

While the multi-actor approach is critical to make TOD work, there is also a need for a multi-function approach, whereby the transportation functions are integrated with other types of development. The multi-actor and multi-function approaches must be integrated, a need that can be met in a PPP, especially in the case of organizing TOD. TOD is a strategy that reveals itself in complex urban (re)development projects, making PPPs relevant. The role of joint-venture PPPs is described in Chapter 6.

Table 7.2   Transit-oriented development actors and their goals (Belzer et al., 2002a).

Actor	Possible goals
Transit agency	Maximize monetary return on land Maximize passenger numbers Capture value in the long term
Passengers	Create/maintain high level of parking Improve transit service and station access Increase mobility choices Develop convenient mix of uses near station
Neighbours	Maintain/increase property values Minimize traffic impact Increase mobility choices Improve access to transit, services, jobs Enhance neighbourhood liveability Foster redevelopment
Local government	Maximize tax revenues Foster economic vitality Please constituents Redevelop underutilized land
Federal government	Protect 'public interest' and set limits on how federally-funded investments can be used
Developer/lender	Maximize return on investment Minimize risk, complexity Ensure value in long term

7.3.3 MAX light rail project: Portland, Oregon

Metropolitan Area Express (MAX) is a 44 mile light rail system with 64 stations that runs east and west from Portland and connects the cities of Gresham, Beaverton and Hillsboro. MAX is part of an integrated light rail and bus system that serves the urbanized portion of the three counties in the Portland metropolitan area. The system was built in four segments:

(1)   Eastside MAX Blue Line, opened in 1986, stretches 15 miles eastward to Gresham
(2)   Westside MAX Blue Line, opened in 1998, runs 18 miles west to Hillsboro
(3)   Airport MAX Red Line, opened in September 2001, runs 5.5 miles northwest from Gateway Transit Center to PDX airport
(4)   Interstate MAX Yellow Line, opened in May 2004, is a 5.8 mile segment that connects the Expo Center in North Portland with downtown and the rest of the transit system.

Portland's light rail network is operated by the Tri-County Metropolitan Transportation District of Oregon (Tri-Met) which was charged with devising a suitable light rail network to attract new development without bringing a disproportionate growth in road traffic. The overall approval rating for the MAX line has been around 90% during the past few years. Eighty per cent of Tri-Met passengers have a car, but choose to use MAX or the bus. Between 1990 and 2001, passenger numbers on Tri-Met doubled from 45 million to just under 90 million. Westside MAX also has a new maintenance depot, at SW 170th station, consisting of a two storey workshop which can service six carriages simultaneously. Investment worth US$2.4 billion has occurred along the MAX line since the decision to build was made. In contrast to Eastside MAX, Westside MAX travels through stretches of undeveloped land, as well as the cities of Beaverton and Hillsboro; the line has become a magnet, attracting nearly 7000 housing units and more than US$500 million in new transit-oriented communities within an easy walk of the stations.

The first phase, Eastside MAX, connected Gresham, Oregon State's fourth largest city, with downtown Portland, and was constructed between 1982 and 1986. The Hillsboro to Portland Westside MAX light rail system began service in September 1998, construction having begun on the 18 mile light rail line in May 1994. Initial planning for the Westside was completed in 1979, with a tentative route (following the Sunset Highway) selected in 1983. The MAX system was expanded to 38 miles with the opening of Airport MAX in September 2001. Airport MAX provides a direct service between Portland International Airport and downtown Portland. The Yellow Line includes ten new stations between Expo Center and Rose Quarter in North Portland, with trains travelling through downtown Portland.

Following the opening of the Yellow Line, the next project is the 6.5 mile Green Line (south corridor), for which preliminary engineering has started, with construction due to start in 2006 and operations in 2009 (www.railway-technology.com/; movingtoportland.net/).

Joint venture

In 1998, a PPP was formed between three local government agencies representing the City of Portland (Tri-Met, Port of Portland and Portland Development Commission) and Bechtel Enterprises to extend the region's award-winning MAX light rail line to Portland International Airport (PDX), which opened on 10 September 2001.

Bechtel Enterprises contributed US$28.2 million towards the US$125 million project and in return, in partnership with Trammell Crow, received rights to develop Cascade Station, a 120 acre (49 hectares) TOD at the Portland International Center (PIC) business park near the airport entrance, forming Cascade Station Development Company to do so (www.cfte.org). The three public partners contributed US$45.5 million (Tri-Met), US$23.8 million (City of Portland) and US$28.3 million (Port of Portland). The partnership required open communication, a significant amount of due diligence and a number of interlocking agreements to protect the public investment (www.trimet.org). The land around the MAX Red Line will be developed under a single masterplan, creating around 10 000 jobs and US$400 million worth of hotels, conference facilities, restaurants, retail, entertainment and office space served by two light rail stations. Completion is scheduled for 2015.

7.4 PPP in complex urban area development

Step three in the twelve steps to urban (re)development, is to 'forge a healthy private/public partnership' (Leinberger, 2005). In general, most successful urban developments are PPPs, but PPP does not *per se* lead to successful development. Certain necessary preconditions must be met to make PPPs in area development succeed (see Chapters 2 and 6). Some general lessons that can be learned from the joint venture PPP examples in Section 7.3 are described below.

7.4.1 Creating joint venture PPPs

Urban area (re)development projects are designed to solve problems concerning liveability and accessibility. Public–private partnerships in complex urban area development mostly take the form of joint ventures (further explained in Chapter 6). This kind of PPP involves both a multi-function (the integration of housing, retail, office, open space) and a multi-actor (government, developers, neighbours) approach. The application of both approaches adds a layer of complexity because of the typically horizontal structure and the diversity of content. Here, the private and public entities give direction and share the risks inherent to the process. The added value of joint venture PPPs is the higher quality solution, the adjusted goals and interests, the synchronization with adjacent developments and the dispersal of risk.

In urban redevelopment PPPs, private actors are often involved from the beginning. Naturally oriented around issues of development and exploration, private actors need to be involved from the early stages of a project to maximize both their contribution and their gain from being involved in the larger process.

7.4.2 Social and commercial benefits

For PPPs to work, both the public and private sectors must reap benefits from the partnership, here the benefits being commercial profits and social returns on investment. From urban redevelopment such benefits can be improved environments, strengthened community services, growth in minority enterprises, employment and economic growth. Most successful examples of PPPs in city redevelopment are downtown economic developments with social spin-off.

Commercial profits in TOD are derived primarily from property assets that are part of the transport system development (Freilich, 1998; Capitol Region Council of Governments, 2002). Techniques employed to generate profits may include (Freilich, 1998):

- Property taxes or assessments
- Excess land acquisition, such as land and air rights leasing
- Negotiated private sector investments in property and transit station capital costs
- Connection fees for direction tie-ins to transit stations and/or
- Concessions at transit stations.

The majority of studies support the position that property values, property tax revenues and land values increase significantly when located near rapid transit stations (Cervero 1994; Economic Research Associates, 1995; Zykofsky, 1998/1999; NEORail II, 2001). According to Cervero (1994), joint development projects add more than US$3 per square foot to annual office rents. Cervero's research also showed that,

> '. . . office vacancy rates were lower, average building densities higher, and shares of regional growth larger in station areas with joint development projects. Combining transit investments with private real estate projects appears to strengthen these effects.' (Cervero, 1994)

There are several ways that TOD investments bring social benefits to the surrounding areas. Ultimately, TODs address the economic decay prevalent in some urban neighbourhoods. By reinforcing community centres, neighbourhood reinvestment and economic development activities, TOD projects buttress the revitalization of urban neighbourhoods and centres. Throughout the USA, in a variety of different contexts, TODs have transformed communities. As an example, Zykofsky (1998/1999) reported,

> 'Several cities in the San Francisco Bay Area have replaced blighted sections of their community with new residential and commercial development close to transit. The City of Richmond transformed a deteriorated park in its downtown, just one block from a BART station, into a retail and residential centre. Anchored by a supermarket and drug store, the 78 000 square-foot centre includes several neighbourhood-serving shops that combine to create 200 new permanent jobs.'

Introducing housing options within TODs helps lower income families live without dependence on a car, which in 1995 could cost families up to US$7405 per year (American Automobile Association, 1995). TODs also give pedestrians more transit options, which is especially helpful for children, the elderly and those with disabilities.

7.4.3 Performance

Although PPPs in urban area development can open up real opportunities, joint development can be challenging. The dominant fear concerning urban area development is that private sector involvement will lead to private cherry picking of projects and an increased demand for public investment. In reality, partnerships in urban area development can prevent private skimming of profits through the intense public decision making that takes place and through co-operative planning and development. Although organized as joint ventures, the public sector is able to maintain control over supply, quality and prices in PPPs. Generally, the greatest challenges in using PPP in urban area developments relate to the problems emerging from the multi-actor and multi-function approaches (Lyall, 1982; Belzer & Autler, 2002a; Bult-Spiering et al., 2005; www.newurbanism.org).

Resolving the problems inherent in the multi-function approach is necessary. Combining different functions necessitates careful co-ordination of activities; links between different functions and the larger issues of accessibility and liveability must be documented in a system-wide plan in which the intersections are clearly defined. These plans need to be comprehensive and flexible, allowing for optimization of the project over time.

The multi-actor approach can create three main problems: responsibility, scale and language. As many are involved, few feel responsible for the project and its outcomes. Each actor may operate on a different scale, from local to regional; therefore, it can be difficult to engage actors working with different geographic orientations. As each actor comes from a different organizational culture, it is possible that each may bring their own language into the project. Agreeing on functional definitions, goals and preferences can be a critical part of a PPP; to ensure success, it is imperative that the outputs of the larger project are clearly defined and understood by all participants. Each urban project, depending on the type of area where it is located and the context within which it sits, will have very different strategies and outputs. Therefore, for each project type, the performance criteria for each actor need to be explicit. Crucial in the PPP process is the stability of equal interpersonal relationships, a stable local political leadership over a long time period and a common sense of urgency.

7.5 Summary

In this chapter, two examples of joint-venture PPPs in the USA have been described: Baltimore's inner-city redevelopment and the MAX light rail project in Portland. Both cases were placed in the context of PPPs in urban development in the USA. Joint-venture PPPs offer the opportunity for both the public and the private sector to improve an urban area's long term quality using both public and private revenues. However, several problems can arise from the required multi-function and multi-actor approach. Therefore, comprehensive and flexible plans must be developed, and the actors' differences in responsibility, scale and language need to be resolved.

References

American Automobile Association (1995) *Your Driving Costs* (pp. 4–5). AAA.

Belzer, D. & Autler, G. (2002a) *Transit-oriented development: moving from rhetoric to reality*. Discussion paper prepared for The Brookings Institution Center on Urban and Metropolitan Policy and the Great American Station Foundation.

Belzer, D. & Autler, G. (2002b) Countering sprawl with transit-oriented development. *Issues in Science and Technology*, Fall 2002, 51–58.

Bult-Spiering, M., Blanken, A. & Dewulf, G.P.M.R. (2005) *Handboek Publiek–Private Samenwerking*. Utrecht: Lemma.

Burchell, R.W., Listokin, D. & Galley, C.C. (2000) Smarth growth: more than a ghost of urban policy past, less than a bold new horizon. *Housing Policy Debate*, 11 (4), 821–879.

Capitol Region Council of Governments (2002) *Liveable Communities Toolkit: A Best Practices Manual For Metropolitan Regions, Tools for Towns*. Hartford, CT: Capitol Regional Council of Governments.

Cervero, R. (1994) Rail transit and joint development: land market impacts in Washington, D.C. and Atlanta. *Journal of the American Planning Association*, 60 (1) 83–94.

Cervero, R. (1998) *The Transit Metropolis: A Global Inquiry*. Washington DC: Island Press.

Cox, W. & Alm, R. (2002) Off the books: the benefits of free enterprise that economic statistics miss. *Reason Magazine*, August 2002.

Economic Research Associates (1995) *Transit Case Studies for the City of Hillsboro LRT Station Area Study*. Los Angeles: Economic Research Associates.

Ewing, R. (1997) Is Los Angeles-style sprawl desirable? *Journal of the American Planning Association*, 63 (1), 107–125.

Fosler, R.S. & Berger, R.A. (1982) *Public–Private Partnership in American Cities: Seven Case Studies*. Lexington, MA: Lexington Books.

Freilich, R.H. (1998) The land-use implications of TOD. *Urban Lawyer*, 30 (3) 547–572.

Greater Baltimore Committee (2005) *Annual Report 2004–05 Anniversary Commemorative Edition*. Baltimore: Greater Baltimore Committee.

Janssen-Jansen, L.B. (2004) *Regio's uitgedaagd, 'Growth Management' ter inspiratie voor nieuwe paden van pro-actieve ruimtelijke planning*. Assen: Van Gorcum (dissertation).

Knaap, G. (2002) Talking smart in the United States. Paper prepared for the *International Meeting on Multiple Intensive Land Use*. Gouda: Habiforum.

Leinberger, C.B. (2005) Turning around downtown: twelve steps to revitalization. *Brookings Institution Research Brief*, March 2005, 1–23.

Linder, S.H. (1999) Coming to terms with the public–private partnership. *American Behavioral Scientist*, 43 (1), 35–51.

Lyall, K.C. (1982) A bicycle built-for-two: public–private partnership in Baltimore. In: R.S. Fosler & R.A. Berger (Eds). *Public-Private Partnership in American Cities* (pp. 17–58). Lexington, MA: Lexington Books.

Lyall, K.C. (1986) Public–private partnerships in the Carter years. In: P. Davis (Ed) *Public–Private Partnerships: Improving Urban Life* (pp. 4–13). New York: Academy of Political Science.

NEORail II (2001) *The Effect of Rail Transit on Property Value: A Summary of Studies*. Cleveland, OH: NEORail II.

Newman, P.G. & Kenworthy, J.R. (1996) The land use-transport connection: an overview. *Land Use Policy*, **13** (1), 1–22.

Rubin, J.S. & Stankiewicz, G.M. (2001) The Los Angeles Community Development Bank: the possible pitfalls of public–private partnerships. *Journal of Urban Affairs*, **23** (2), 133–153.

US Department of Transportation, Federal Transit Administration (1995) *FTA Fiscal Year 1996: Apportionments and Allocations.* Washington DC: US Department of Transportation.

Zykofsky, P. (1998/1999) Why build near transit? *Transit California,* December 1998/January 1999.

Further Reading

Austin, J. & McCaffrey, A. (2002) Business leadership coalitions and public–private partnerships in American cities: a business perspective on regime theory. *Journal of Urban Affairs*, **24** (1), 35–54.

Bernick, M.S. & Freilich, A.E. (1998) Transit villages and transit-based development: the rules are becoming more flexible – how government can work with the private sector to make it happen. *Urban Lawyer*, **30** (1).

Cervero, R. (2000) Growing smart by linking transportation and urban development. *Built Environment*, **29** (1), 66–78.

Cervero, R., Ferrell, C. & Murphy, S. (2002) Transit-oriented development and joint development in the United States: a literature review. *Transportation Research Board: Research Results Digest*, **52**, 1–144.

Davis, P. (Ed) (1986) *Public-Private Partnerships: Improving Urban Life.* New York: Academy of Political Science.

Dunn, J.A., Jr (1999) Transportation: policy-level partnerships and project-based partnerships, *American Behavioral Scientist*, **43** (1), 92–106.

Dunphy, R., Myerson, D. & Pawlukiewicz, M. (2003) *Ten Principles for Successful Development around Transit.* Washington DC: Urban Land Institute

Landers, J. (2002) MAX Transit. *Civil Engineering*, **72** (1), 44–49.

Newman, P.W.G. & Kenworthy, J.R. (1999) *Sustainability and Cities: Overcoming Automobile Dependence.* Washington DC: Island Press.

Tumlin, J. & Millard-Ball, A. (2003) How to make transit-oriented development work. *Planning*, **69** (5), 14–19.

Webliography

Baltimore Regional Partnership www.balto-region-partners.org/ (accessed June 2005)

Center for Transportation Excellence www.cfte.org (accessed April 2005)

Congress for the New Urbanism www.cnu.org/ (accessed December 2004)

Greater Baltimore Committee www.gbc.org/ (accessed June 2005)

Moving to Portland movingtoportland.net/public_transportation.htm (accessed April 2005)

New Urbanism www.newurbanism.org/ (accessed December 2004)

Railway Technology www.railway-technology.com/projects/portland/ (accessed April 2005)

Smart Growth America www.smartgrowthamerica.com/ (accessed December 2004)

Transit Oriented Development www.transitorienteddevelopment.org/ (accessed December 2004)

Transit Oriented Development Advocate www.todadvocate.com/ (accessed December 2004)

Tri-Met www.trimet.org (accessed April 2005)

TDM Encyclopedia www.vtpi.org/tdm/tdm45.htm (accessed December 2004)

8 The Future of Public–Private Partnerships

Co-author: Anneloes Blanken

All over the world we have seen public–private partnerships (PPPs) emerging in urban development, for the provision of transport infrastructure, schools, hospitals and other public services. For years, PPPs between governments and private organizations have formed a well-established means of providing facilities that governments have neither the resources nor the expertise to supply alone (Forsyth, 2004). As mentioned in the preceding chapters, various types of or concepts for partnerships exist: concession contracts are the most common worldwide.

Despite the continuous application and development of PPP schemes, the arrangements have been criticized by both the public and private sectors, and PPP is a constant topic of political debate. A major criticism concerns transaction costs and the time spent on PPP procurement being disproportionately high: many procurement routes could take up to 18 months and even longer before financial closure. Another aspect of PPPs that is often criticized, and especially applies for concessions, is the contracting out of services instead of 'real' partnerships on the basis of mutual responsibilities and risk sharing.

Hence, in recent years, many efforts have been undertaken to improve the performance of PPPs, both at the operational and strategic levels. At the operational level, policy measures have been implemented to streamline the procurement process and to make the functioning of PPPs more efficient. At the strategic level, new forms of partnerships have been developed. This development can be characterized as the emergence of portfolio partnerships instead of single facility or mono-function partnerships. Trends and developments will be discussed in this chapter.

8.1 Lessons

The introduction of PPPs was accompanied by several expectations. In this book we have discussed various types of PPPs and conclude that the reasons, motives and added value underlying the start-up of a partnership differ depending on the type of PPP. Sharing risks and revenues, for example, is among the primary motives for introducing joint ventures; transferring risks and revenues is among the motives for concession arrangements. Both types of PPP, however, are seen as an inevitable reaction to the blurring of boundaries between the public and private sectors, and to the unfavourable financial positions and severe spatial problems that governments in several countries have been dealing with. Furthermore, PPPs were meant to provide certain benefits in terms of efficiency and risk reduction above the traditional methods of procurement.

8.1.1 Expectations

The expectations that came with the launch of PPPs had mainly to do with the sources of efficiency and were accompanied by the introduction of a role for the private sector in providing public services.

Concentration on core activities and steering

One of the efficiency gains accompanying concession PPPs is that contracting with an external private provider saves the former service provider, i.e. the public actor, time and effort. This will allow the public sector organization to focus and specialize in so-called 'core' activities and thus maximize some measure of efficiency (Grimshaw et al., 2002). In joint venture PPPs, the public agent remains heavily involved during the whole process, but shares the effort with the private partner(s).

Market discipline

The discipline of the marketplace is also said to be an important device in driving down the costs of construction and services provision. The private sector is expected to be more efficient and cost-effective than the public sector (Grimshaw et al., 2002). Concession arrangements are viewed as a means of reducing, if not eliminating, unnecessary sources of waste, inefficiency and ineffectiveness from the value chain by replacing arm's-length and destructive forms of rivalry and competition with productive forms of collaboration and learning. Joint ventures are launched to find an optimal balance between public interest and market discipline.

Output specification

Concession arrangements are embedded in formally specified contracts. This is believed to improve control and monitoring processes on the levels of both quality and output. Traditional forms of procurement are mostly open-ended contracts and less accommodated to steering and monitoring. In this respect, efficiency gains are expected through specification of outputs to be delivered rather than the detailed specification of how they are to be produced (Glaister, 1999). In addition, periodic renewal of the contract, which is assimilated in the concession contract in several countries, is seen as a competitive incentive to induce higher motivation and productivity among management and workers providing the service.

Private sector expertise and flexibility

Concession arrangements are, according to Grimshaw et al. (2002), championed in the name of increased speed, flexibility, integration and innovation by blurring the organizational boundaries between public and private organizations. Besides, the involvement of a private service provider is believed to be critical for enhancing the flexible capacity to adapt and respond to new pressures and conditions, as well as the potential for innovation in actual service provision. In joint venture PPPs, as we described in the earlier chapters, it has been argued that market knowledge is being used to increase the quality of the built environment and improve the integration of various functions. Furthermore, joint ventures are established to align the (urban) development with the wishes and requirements of the marketplace.

8.1.2 Value for money

The gains in efficiency and/or improved service delivery of PPPs compared with traditional procurement practices can be summarized in the term 'value for money'. Value for money focuses not only on cost-effectiveness in the delivery of public services, it also appraises the value or quality of the delivery; it is the optimum combination of cost and quality in meeting the needs of service users (Office of Government Commerce, 2000). Thus, the argument for PPP arrangements is that they offer the potential to secure better value for money and greater innovation in the delivery of public services, with a focus on both efficiency and quality (Institute for Public Policy Research, 2001). In the UK and some other countries, value for money is explicitly required for proceeding with a concession PPP project. Value for money assessment is then determined (*a priori*) by a public sector comparator (PSC).

In accordance with various studies, we conclude that PPPs fall short of expectations. Although it has been hard to establish proper comparisons with what a public sector alternative would deliver, results show that concession arrangements, in particular, are delivering only marginal savings (Institute for Public Policy Research, 2001). Cost savings between the concession and public option depend almost entirely on the aspects that are included in the transfer of risk to the private sector and how these have been valued (Institute for Public Policy Research, 2001). Their report into PPPs (Institute for Public Policy Research, 2001) looked at all the publicly available evidence comparing bid costs using PSCs: the results were mixed. Some projects for roads and prisons showed significant cost savings in terms of prescribed methodology (National Audit Office, 1999a), although less than the public agencies had estimated (National Audit Office, 1998, 1999b), but other PFI schemes demonstrated more marginal savings of between 2% and 4% (Maltby, 2003). In particular, the health and education sectors gave cause for concern in this respect. Research indicated specifically that the amount of risk transferred to the private sector in health concession contracts was almost exactly the amount required to bridge the gap between the cost of concessions and conventional procurement. This suggests that the function of risk transfer was to disguise the true cost of concessions (Edwards & Shaoul, 2003).

Other researchers also demonstrated similar results in the education sector (Ball et al., 2001; McCabe et al., 2001). Audit Scotland (2002) released data about operational PFI schools and dismissed any final cost savings in PFI as 'narrow'. In addition, the Audit Commission (2003) compared 12 traditionally funded schools with 17 PFI schools and found that the quality of PFI schools was not as good as schools built by more traditional means, while the best examples of innovation came from traditional schools. Also, according to Ball et al. (2001), value for money for concession arrangements is still uncertain, mainly because of the lack of innovation.

In addition, there is widespread concern, both in the private and public sectors, that the process by which the value-for-money assessment for a particular project is determined has, in many cases, been highly problematic and lacking in transparency.

In studying joint-venture PPPs it is even harder to define the value for money. The advantages of joint ventures for both public and private sector are more 'intangible', e.g. higher urban quality and better co-operation. Sociological aspects, such as trust and commitment, are important determinants for the success of joint venture PPPs.

In general, we are able to conclude that different types of PPPs, especially concession arrangements, do not come up to expectations. In preceding chapters we have analysed conditions for the success of PPP projects. These conditions can be categorized into those concerning context, organization and project levels. For each level we can formulate lessons for the future, based on the analyses in preceding chapters and comparison between the various types of partnerships.

8.1.3 General lessons

The success of PPP does not depend on one single aspect. Hence, the following list of lessons should be considered integrally. In any case, PPP should be a strategic choice: PPP is not a goal in itself but a method by which to attain certain objectives. We will discuss this issue later on.

Changes required at context level

Legislation should be adapted to make optimal use of PPPs for the provision of public goods and services. This entails:

- Establishing or clarifying the legitimacy and powers of public authorities to enter PPP contracts
- Integration of different spatial procedures that apply for construction projects

■ Removing tax anomalies that can weigh against concession approaches
■ Refining public expenditure capital control regimes to accommodate PPPs.

In addition, well-defined and transparent procurement strategies and standardization of procedures are needed. Issues or problems that have been encountered during the development of PPP projects are related to poorly defined procurement methodologies and lack of standardization; this results in poorly structured contracts, lengthy and costly development and procurement processes, high bidding costs and reduced value for money. In Europe, governments are obliged to award contracts through a process of competitive tendering.

Finally, a steady flow of projects and potential for generating cash flow are required. In The Netherlands, Germany and many other countries, the application of PPP varies in time and space; as a result, it is uncertain whether PPP is a structural strategy or not, which is an important issue for private sector organizations to make well-founded investment decisions. In addition, the generation of a project flow of PPP projects is of interest to private parties with respect to building expertise and allocating project risks. Governments can issue some form of commitment to the PPP approach by generating a continuous stream of projects and possibilities for value capture and balancing costs and revenues.

Changes needed at organizational level

Professionalism of the public partner is required; developing capabilities on the public side is an important condition for success. Ahadzi & Bowles (2003) concluded from a survey among public and private managers involved in PPPs that on the public side the following capabilities are important:

■ Technical capabilities, such as being able to establish project parameters and clear output specifications
■ Organizational capabilities, such as commitment and the level of collaboration within the public sector.

This last conclusion is in accord with findings of a study in The Netherlands (Smit et al., 2002) indicating that public tensions are the major reason for malfunctioning of a PPP. Other issues mentioned by Ahadzi & Bowles (2003) are the ability to sensitize public opinions effectively, and the ability to develop clear evaluation criteria. A key factor in professionalism is the establishment of a central PPP knowledge unit, as has been done in many developed countries. This unit should be sufficiently embedded in all government departments involved in urban development and infrastructure projects. The influence of the PPP unit is usually most intense in the early stages of the PPP programme, and will decline over time as policy and procedures are established within the public sector, and as capabilities of the departments to undertake PPP procurements are further developed. Another influence is the level of existing experience with PPPs. The more projects the public actor procures, the more experience it will obtain.

Scale is needed in terms of projects and also in terms of private consortia. One of the big advantages of the Spanish marketplace in relation to other countries is the large size of the construction firms. Private sector companies need to be large to justify their bearing substantial risks, and also to obtain and disseminate knowledge and experience in PPPs.

Finally, a stable and confident government policy is required. It appeared important that ministers should be responsible champions and promote the PPP process within a government. In Chapters 6 and 7 on joint ventures, the importance of stable local political relationships and a common sense of urgency was stressed. In relation to this, trust and mutual respect are prerequisites for success. In Spain, for instance, PPP policy seems to be based on trust, whereas in The Netherlands it is based on distrust. It is expected that projects will be performed more efficiently and effectively when trust is integrated into partnering concepts.

Changes needed at project level

Appropriate levels of risk transfer are needed, as described in preceding chapters. An important aspect of PPPs is the understanding of the potential risks associated with a project, and the implications that these risks could have in terms of project costs, timing and quality. The valuation and allocation of risks is a significant part of PPP contracts. Consequently, the valuation of risks has recently been the subject of debate; one of the main topics of discussion has been the fact that some risks are being transferred to the public sector when problems start to arise. It appears that the commitment of the private sector in accepting an appropriate degree of risk (without public sector guarantees) is essential in PPP projects.

The Mott MacDonald (2002) study (see Chapter 4) indicated that the most important stage is the development of the business case, which encompasses the definition of benefits, requirements and scope of the works. It is not surprising then that Ahadzi & Bowles (2004) concluded that the appointment of a dedicated bid manager and involving all relevant stakeholders at an early stage have a major positive influence on negotiations in the tendering process. These two attributes are needed to be able to define an optimal business case.

Early involvement of the key stakeholders leads to knowledge exchange between partners and consequently to integration of knowledge in the project. In Chapter 6 it was indicated that early involvement of the private sector creates more opportunities for establishing added value in plans and realization, as well as a decline in risks, since realization and exploitation are influenced by decisions taken early in the process.

As already mentioned, PPP projects, especially concessions, appear to be more successful and generate more value for money when clear and measurable output specifications are formulated instead of input specifications. The use of output specification results is believed to be an incentive for improved monitoring, which may increase value for money. If the preliminary work of the public partner is too detailed and already embedded in tendering documents, space for innovation may be limited; an important motive for PPPs, however, is the expected process and product innovation. Nonetheless, it is important that approvals and permits should be arranged before the start of the project. When this has not been done properly, it may cause delays and extra costs for the public partner. Better value for money can be achieved by focusing on added value instead of solely financial aspects of the contract through the following incentives:

- Linking an element of contractual payments to the tangible benefits brought to users: finding a job, passing exams, or achieving improvements in standards of health
- Linking an element of the payment stream to measures of user satisfaction

Although some incentives are now used in some countries, the extent to which these should be included within deals is a key issue which should be thoroughly considered.

Clear division of responsibilities and roles and good co-ordination between the different partners will improve both the process and the final product. In PPP projects the definition of the roles of public and private partners in the financial domain and in the delivery of public services is an important issue. In contrast to privatization, in PPP projects it is important that the public sector retains control of core areas of responsibility. To guarantee the success of PPPs, monitoring, control and evaluation are essential throughout the entire process, because they will enable the project to be steered in the desired direction permanently.

8.2 Scope of PPPs

Today, PPPs are applied in a wide variety of sectors. The political debate and the media are now concentrating the discussion on whether so-called public or collective goods can be provided by the private sector. Some of these goods are so desirable that no charge has to be paid for the use of collective goods. Savas (2000) defines these goods as 'worthy goods'. Since the origin of the welfare state, many of these goods have been directly produced by government. Whether collective goods are seen as worthy depends on cultural-historical and economic developments. In Spain, for instance, it is widely accepted that the public pays for the use of roads, while in The Netherlands toll roads have been politically unacceptable. The history of concessions in infrastructure also indicates how political debates may change over time. In Portugal, for instance, a country with a long history of central government control, concessions have boomed in a very short period of time.

In transport infrastructure, political acceptance for PPPs is becoming widespread, but this is still not the case, in other sectors, particularly education, healthcare and social services. This is not surprising because generating scale, defining clear output specifications, defining the risks involved, setting clear responsibilities and other measures mentioned are easier to implement in PPPs for the provision of economic infrastructure than they are in the healthcare and education sector, the so-called social infrastructure sector. Several characteristics of the social infrastructure can be identified that raise problems for implementing PPP contracts:

- *The ambiguous nature of services (Kirkpatrick, 1999)* In some social infrastructure sectors, such as healthcare and education, problems are revealed in drawing up output-based measures in PPP arrangements, due to the constantly changing nature of the services needed. Clinical outcomes are at present extremely difficult to define and are subject to a long term health agenda. The difficulties in detailed descriptions about the performance to be provided led to problems of defining activity and output, identifying unit costs and monitoring compliance. Public as well as private organizations encountered major conceptual and methodological problems in detailing the causal relationship between action and effect, and linking this with needs and objectives (Kirkpatrick, 1999). The complexity of services has resulted in increased transaction costs which may exceed efficiency gains, as attempts are made to specify and enforce detailed contingent claims contracts (Boyne, 1998).
- *Increased difficulty of service co-ordination (Kirkpatrick, 1999)* The formal division between purchaser and provider has led to 'marketized' relationships which encourage competitive and combative forms of behaviour, and prioritize immediate results rather than development (Kirkpatrick, 1999). In some cases this has created a them-against-us approach in relations between public and private actors. In his study on local authority managers, Walsh (1995) concluded that contracting hampers co-operation processes between organizations and that it fosters organizational conflict, which could make authorities less flexible.
- *Relatively small scale of social infrastructure projects* The procurement process for PPP projects in healthcare and education is of comparable length to that of major capital schemes. Small schemes typically face transaction and bid costs similar to those of major capital schemes (HM Treasury, 2003). However, concession arrangements need a specific volume in terms of concession period and costs to recover these expenses. This indicates that, in relation to the level of capital investment undertaken by the schemes, procurement times are disproportionately long and procurement costs disproportionately high, which makes it difficult for small PPP schemes to consistently deliver value for money
- *The separating out of ancillary and core services (Institute for Public Policy Research, 2001)* In the case of a prison, for example, it is not possible for the private sector partner to integrate thoroughly the design and construction of the building with its operational phase and make productivity gains through the way it manages the single most important input in any public service, the workforce. In the prison service, a full range of services (both core and ancillary) can be procured, including the prison management and their staff.

By contrast, in healthcare and education a very restricted model of concession arrangements is being used, with the operating element encompassing only a narrow range of ancillary services such as maintenance, cleaning and occasionally information technology (Institute for Public Policy Research, 2001); the main labour intensive forms of service provision are excluded and the possible gains from greater efficiency are therefore limited.

■ *Lack of skills and capabilities* In the health and education sectors there is not one purchaser who has over time built up expertise in contracting, but a number of potential purchasers, because some responsibilities and tasks have been transferred to local authorities. For prisons, the Prison Service is the sole purchaser and has, over time, built up expertise in contracting; arguably, the Highways Agency has developed similar expertise in relation to contracting for PPP road projects. This key feature is absent in education and health (Institute for Public Policy Research, 2001). Particularly in the education sector, in which two public agencies (the local education authority and the board of governors of the school) are mutually responsible for the PPP process, are problems observed in relation to an already complex appraisal methodology and process, that limit the value of the technique and PPP regulation (Edwards & Shaoul, 2003).

To overcome these deficiencies, it was suggested that alternative PPP schemes should be developed which would be convenient for the provision of social infrastructure. Recently, such schemes have been designed to allow greater participation of all non-governmental actors in shaping development policy. Consequently, in several countries a transition is taking place from concession arrangements to more innovative schemes of PPPs in which social infrastructure projects are bundled together. These portfolio strategies will be discussed in Section 8.3.

8.3 Portfolio strategies

In portfolio strategies, solutions are found at the cutting edge of mutual responsibilities for strategic planning and the operation and maintenance of products for public and private actors, in order to deliver optimal added value. In some sectors, such as schools and healthcare, in which the capital value of individual projects is small, the UK government is using these new partnership methods to deliver value for money, examples being building schools for the future (BSF) in the education sector and local improvement finance trusts (LIFTs) in healthcare. In Germany also, initiatives are being taken in the education sector to achieve economies of scale by combining several school projects in PPP programmes.

In portfolio partnerships, besides bundling several infrastructure projects together, the partners agree upon a long term co-operation in which the overall contract time is independent of the duration of the individual contracts: the aim is to align objectives to deliver high efficiency and effectiveness. These new procurement vehicles are expected to improve the speed and reduce the costs of concession procurement.

8.3.1 Programmes

Programmes of PPPs have been developed in several countries to achieve economies of scale. In these partnerships, the demand for standardization and economies of scale has encouraged the 'bundling' of schemes across services such as housing, schools and social services into single larger packages. From that perspective the programmes can be seen as an interim step towards regional portfolio partnerships chosen for integrated co-operation.

The aim of programmes of PPPs is to standardize procedures, reduce the procurement period and lower the high costs of bidding for contracts. This arrangement makes subsequent contracts easier and cheaper to secure, and further integrates public sector provision into the marketplace (Rutherford, 2003).

The organization of the partnerships in portfolio programmes is adjusted to obviate the deficiencies in some specific social infrastructure sectors that have been complicating delivering value for money by using concession arrangements (see Section 8.2.1). In PPP programmes, the relatively small scale of projects in the sectors in question are neutralized by bundling several types of properties into one contract structure, thus permitting transaction costs to be distributed amongst them. In addition, efficiency is expected to be increased, while expertise in contracting can be built up among the public organizations involved which also can improve the speed of procurement.

A number of examples can be given in this respect. In Germany, 15 schools in Meschede (nine elementary, three middle, one secondary modern, one grammar and one special) and their sports facilities required refurbishment. While some of these schools formed an organizational entity or were combined in joint locations, a total of 16 buildings had to be refurbished. A PPP viability study predicted efficiency gains of between 2% and 15% compared to traditional procurement by bundling these facilities into one project structure.

In Monheim, all 13 council-run schools (eight primary, one middle, one secondary modern, one comprehensive, one special and one grammar), together with sports facilities and gymnasiums, are being refurbished using a step-by-step pooling scheme. In addition, in Witten, a PPP model scheme has been launched to refurbish and extend two grammar schools (€13 million). The project involves the amalgamation of two grammar schools at an inner-city site and the relocation of a secondary school to the premises vacated by the grammar school; comprehensive refurbishment and extension is required at both sites. An EU-wide invitation to tender was issued in November 2003 and contracts were signed in August 2004. The expected efficiency gains are estimated at 9.3% over conventional procurement.

Similar developments are occurring in the UK. In Glasgow, a large 'bundled' PPP scheme for 29 secondary schools was put out to tender. It was claimed that, following completion of the capital investment programme, efficiency savings of up to 17.5% on facilities management costs would be achieved. It was also considered that this option would lead to uniformity of service provision to all schools as the service charges would not be competing against other claims on the capital budget. Likewise, in a Falkirk project, several school properties were bundled together: the contract was for the provision of five schools and entailed the provision, operation and maintenance of all schools, including information and communication technology, catering and janitorial services. Like the other programmes mentioned, the arrangement did not include responsibility for teaching or learning.

Although the initiative is full of promise, a potential hazard accompanies portfolio programmes in that there is a danger that the public procurer will be 'locked in' to a single service provider over a long period of time. This may lead to a lack of pressure for quality improvements and the possibility that a service provider could get away with coasting rather than striving for continual improvement. The removal of the possibility of changing providers within the lifespan of a PPP contract runs counter to the argument for retaining the market-place, i.e. efficiency is gained by a degree of contestability in service provision. One-off contestability at the beginning of a contract is unlikely to make up for the 25–30 year period in which a single provider has an effective monopoly. Of course, benchmarking and payment deductions can help prevent under-performance.

8.3.2 Regions

Recently, the UK government has developed innovative approaches towards PPPs in which several collaborative incentives are involved. These are the so-called regional portfolio partnerships. These emerging models of PPPs are putting business between the public and private sectors onto a much sounder basis compared to concession arrangements which have an element of 'them and us'. The concepts have higher ambitions and attempt to solve previously intractable problems, either within the public sector or in the way in which public sector bodies worked with each other or with the private sector.

In these new PPP arrangements, the idea of multi-projects is used by batching different projects together, which increases the scope for co-operative arrangements. Because transaction and development costs and procurement times are disproportionately high for concession arrangements in social infrastructure sectors, value for money can be obtained consistently when small schemes are bundled together.

In some sectors, such as schools and healthcare, the UK government is using these new approaches in PPP arrangements to improve both the quality of the public sector client, and its co-ordination across a sector: a range of new procurement models are being introduced in which there is benefit in bundling projects together and ensuring that the timing of projects maximizes market interest; they are therefore likely to be most applicable where small projects can be grouped together. Such procurement vehicles are promised to improve the speed and reduce the cost of concession arrangements, improving value for money, and allow concessions to be used to provide:

- The LIFT initiative in new areas of healthcare where it had previously been impractical, such as in primary care facilities
- Investment in schools such as those in the 'building schools for the future' programme.

The procurement in regional portfolios is different from concession projects and even more adjusted to the deficiencies involved in health and education than are portfolio programmes. Concession procurement is usually concerned with the delivery of a single facility with an initial construction phase together with on-going related services, while programme portfolios concern a group of identified buildings. Regional portfolios, by contrast, are concerned with another way of bundling projects. The intention is to develop a chain of contracts for the provision of social sector facilities and services. Instead of bundling several types of properties in one contract structure *a priori*, incentives are given for both public and private partners to extend the contract to other properties besides the existing accommodation. Regional portfolios are in that respect incremental strategic partnerships.

In the health and education sectors these partnerships are explicitly geared to be unambiguous in nature. A joint venture company (JVC) will be established in which both public and private partners are involved; this JVC is accountable for the strategic planning and procurement of the facilities in a certain region. This organizational structure allows for flexibility, while local authorities can tailor the project to their local requirements: it is designed to be adaptable (within the parameters of procurement law) to meet evolving local needs of the various participating public bodies over a subsequent period. The basis for these arrangements has been the alignment of partners' broad objectives for collaboration, with joint working and the sharing of some risks implicit or specified in the arrangements. The arrangements seek to create a long term partnering arrangement that brings speed of delivery and value for money. Efficiency gains in terms of bid costs and time are generated by building up the procurement expertise of the JVC and the fact that, to undertake future developments, there is no need to go out to tender again for a bidder. This makes regional portfolio schemes more affordable and quicker to develop.

8.4 Examples of portfolio partnerships

The added value of portfolio partnerships can be illustrated best by the description of some recent examples.

8.4.1 NHS local improvement finance trust

In the UK, the NHS LIFT is an initiative for area-based partnerships in healthcare where concession arrangements had previously been impractical, such as in primary care services (the care received upon first contact with the healthcare system). LIFT can be seen as a tool which contains several partnering elements (4Ps, 2004):

- It is a long term partnership for services accommodation supplied on a 'no service, no fee' basis
- It concerns a joint venture company limited by shares as a local partnership vehicle
- It contains land development aspects (asset and property management planning) for more efficient service delivery
- It links to strategic partnering
- It is a new model for PPPs.

The LIFT initiative has been developed specifically to deliver a step change in primary and social care by developing and supplying new and refurbished health and social care facilities. As a result of the historically inadequate investments in primary care, the quality of primary care buildings can often be assessed as insufficient (National Audit Office, 2005). The LIFT partnership is based on an incremental strategic partnership and is fundamentally about engaging a partner to deliver a stream of accommodation and related services through a supply chain, established according to EU procurement requirements. LIFT offers incentives for both public and private sector participants, as pictured in Table 8.1.

Structure

All LIFT projects are delivered through a common approach, developed by Partnerships for Health (PfH), a national joint venture partnership agency set up by the Department of Health and Partnerships UK.

One of the key benefits of LIFT is that schemes can be bundled together as in programme portfolios. The intention of bundling is that the total capital value of the tranche of schemes will enable economies of scale savings, whilst simultaneously procuring multiple new developments. Under LIFT procurement, there is good residual value because primary care facilities are likely to be used for other activities after their purpose in the LIFT arrangement has been fulfilled. Therefore, the costs of realization, operation and maintenance do not necessarily have to be fully recovered during the lease term. Instead, the LIFT company (LIFT Co) takes the risk that it can recover further capital or income from the property after the initial lease term, which will result in lower periodic payments.

The funding structure of the LIFT model has been designed to allow relatively small capital schemes to be financed in batches (tranches). This allows the LIFT Co to take advantage of the economies of scale. The financing is provided through

Table 8.1   LIFT offers incentives for both public and private sector participants (National Audit Office, 2005). Reproduced with permission of the NAO, London.

Public sector		Private sector	
Incentive	**Benefits**	**Incentive**	**Benefits**
Common approach to procurement	Reduced complexity in negotiations Reduced procurement costs over many projects Ensures more productive use of time	Strategic framework for bringing development opportunities to the private sector	Reduced complexity of negotiating terms Reduced risk of cancellation of projects Contract exclusive to all primary care development in local health economy Reduced risks and costs inherent in new projects
Local stakeholders have a shareholding	Public sector can direct investment in line with local priorities GPs reduce risk of investment through shareholding in portfolio of properties Public sector can reinvest its share of profits in local health	Opportunities to increase income	Increased profitability from LIFT schemes Improved relationships for private sector with local organizations and businesses
Fit for purpose facilities provided	Good quality buildings Improved working environment Fully serviced premises provided	Opportunity to influence local health and estates strategies	New ideas and suggestions will deliver better outcomes and may increase returns
Flexible structure	LIFT model can respond to long term requirements of public sector Shares can be sold/redistributed	Low risk investment	Returns to private sector are reasonable Government backed cash represents increased security
Improved primary care estate	Can extend and co-locate care service Can help meet primary care targets Can secure buy-in from tenants	Facilitates introduction to non-NHS clients	Contact with suppliers of facilities outside the NHS may open up commercial opportunities outside LIFT

debt, with the private sector partner, the NHS primary care trusts and Partnerships for Health (PfH) contributing the equity. The funding structure of a LIFT is shown in Fig. 8.1.

Process

The procurement process of LIFT structures mainly consists of the evaluation of fully budgeted bids of private sector partners for a first tranche of schemes. This gives the public partner the costs and affordability of schemes in the first tranche, as well as setting a benchmark for the costs of future schemes. The procurement

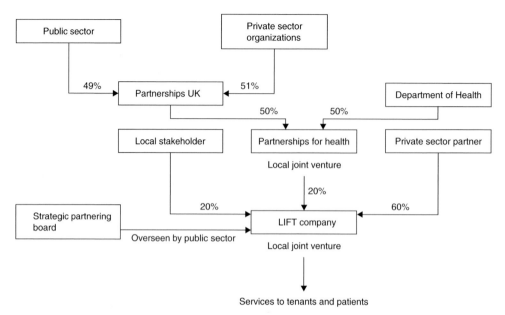

Figure 8.1   Funding structure of a LIFT partnership (National Audit Office, 2005). Reproduced with permission of the NAO, London.

route to select the LIFT partner and supply chain follows an identical path to the process in concession arrangements in the UK, and the standard documentation developed by PfH is closely controlled to prevent unnecessary delays and protracted negotiation over points which have already been agreed elsewhere (4Ps, 2004). When the initial private sector partner has been selected, the purchaser can choose to develop all the tranches in collaboration with this partner and need not procure another partner for the LIFT contract time.

When a local authority has the intention of including facilities within the initial tranche of a LIFT scheme, any relevant land or property is generally expected to be transferred to the LIFT Co at financial closure. Furthermore, when the LIFT Co takes initiatives for new facilities and services to be improved and these involve existing local authority properties, the relevant properties will again be transferred to the LIFT Co (when approved) which will be responsible for obtaining all necessary planning permissions. The LIFT Co will pay a proportion of the profit or a share of anticipated development gain to the local authority. This will be calculated on the increase in value of any properties transferred if planning permission is granted within the years of the transfer of all or part of the property (4Ps, 2004). The tasks and responsibilities of the LIFT Co are mainly related to delivering initial and future services through the supply chain and downstream contracts in which several procurement routes can be chosen.

An important element of LIFT is that stakeholders in a local health and social care community establish a strategic partnering board (SPB), which constitutes a commission where agreement concerning the health and social care needs and requirements to be developed is reached.

Local LIFT Cos are set up as part of the initial procurement process. These are JVCs in which the private sector partner co-operates with health authorities, primary care trusts and local stakeholders to deliver property-related investment, services and facilities to end-users over the period of a decade. A LIFT Co typically delivers developments in tranches through wholly owned special purpose vehicles, which are quite similar to those used in concession arrangements. Usually, the selected private sector partner has a majority shareholding of 60% in the LIFT Co, while other local stakeholders have a minority share holding of 20%. PfH usually accepts the remaining 20% of the LIFT Co equity. The actual co-operation finds expression in strategic partnering agreements (SPAs) and individual lease plus agreements. The SPA sets out the way partners should act together with a view to achieving the objectives of the local LIFT.

As mentioned before, besides the provision of a first tranche of schemes, the LIFT Co is incentivized to be the partner for the provision of future services and facilities in the LIFT area. LIFT Cos themselves are responsible for taking initiatives in which service delivery can be improved. In that respect, exclusivity is a key aspect of the structure of the initiative. Exclusivity means that the LIFT Co will have the exclusive first right of refusal to provide any new facilities and/or services in the region. These initiatives should coincide in the annual agreed business plan and project priorities identified by the public sector participants of the SPB. In addition, the participants must be convinced that the proposals meeting their respective criteria are affordable and the solution proposed by the LIFT Co demonstrates value for money.

Value for money for the public sector is protected in LIFT arrangements, despite the exclusivity granted to the LIFT Co. The LIFT Co is obliged to demonstrate value for money through either benchmarking or market testing in the new projects it initiates, and the other partners in the LIFT have the possibility of entering into contracts with other providers for the services required. The LIFT Co is required to demonstrate that its supply chain delivers good value as well. This will be demonstrated through comparison with current cost trends, benchmarks and market testing with reference to other LIFT projects, both locally and nationally.

The flexibility of the LIFT model is also a key feature in obtaining value for money. The financial structure of LIFT has been developed to be flexible to changes over the length of the partnership. A LIFT Co is not tied into the funder of initial schemes, and the structure allows for financing for each tranche of schemes through separate co-operation arrangements when these give better value for money.

Preliminary results

Initial schemes were focused around deprived inner-city areas where health needs are greatest and prevailing conditions are poorest. The first LIFT developments to be completed were the less challenging ones that could be achieved quickly; later projects are more likely to address LIFT's long term aims. The first 42 LIFT arrangements were approved in England by August 2002; a further nine schemes were announced in November 2004.

Despite problems in getting started, local outcomes were encouraging, as were future prospects for LIFT, although initial performance measurement and accountability frameworks needed strengthening. Most of the developments to date have been well received by local stakeholders, although some proposals have provoked local opposition. In 2005, LIFT is still in its early stages; most LIFT Cos are operational, but few buildings are open. The initial buildings commissioned are likely to be only a fraction of the developments planned under the initiative (National Audit Office, 2005).

8.4.2 Schools for the future

'Building schools for the future' (BSF) is a new UK initiative in the education sector inspired by the LIFT structure for primary healthcare premises. As with the conditions found in primary healthcare premises at the beginning of the new millennium, the capital state of education facilities left much to be desired where quality was concerned. In the initiative, the government has committed itself to a programme of rebuilding and renewal to ensure that all secondary schools in England have facilities to twenty-first century standards (Department for Education and Skills, 2003). The programme focuses on bundling schools into coherent geographical areas and supporting a phased investment programme in the selected area so that educational standards are raised comprehensively for all schools in that area (Partnerships for Schools, 2004).

Structure

In BSF the UK Department for Education and Skills (DfES), Partnerships UK (PUK) and 4Ps have set up a new partnership, named Partnerships for Schools (PfS). Like PfH, PfS will participate with local authorities and the private sector to review, design and implement secondary education provision. The involvement of PfS in BSF schemes can be traced back to the specific procurement expertise PfS will develop and which can be used to procure and negotiate these schemes more effectively and efficiently. The intention of involvement of such a national body is that it will help to solve the problems found in concession procurement which local government, being relatively inexperienced, encounters when negotiating contracts. PfS will be putting up the funds to take concession arrangements beyond its bias towards new-build school projects.

In BSF, a partnership is established to design, build, finance and operate schools, most of which will be extended schools with community facilities such as sports facilities, libraries and adult learning centres. The partnership will

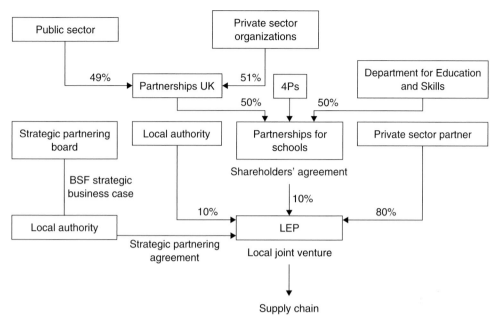

Figure 8.2 Funding structure of an LEP partnership.

concern an agreement between a local authority concerned with education responsibilities in that area, PfS and a private sector partner selected in open competition under EU procurement rules. The resulting partnership is called a local education partnership (LEP). It will be a local JVC focused around the delivery of the strategic investment programme for the region (see Fig. 8.2).

The LEP is responsible for enabling, managing and delivering initial and future projects to participants by way of arrangements with supply chain contractors and contracts in which different procurement routes can be chosen. The LEP will initiate strategic investment plans and act as the co-ordinator for the procurement processes of all facilities and services for secondary education in the indicated area.

The preferential duties of the LEP or exclusivities are embedded in the strategic partnering agreement (SPA), which is an arrangement between the LEP and the local authority. Besides the outline of the exclusivities granted, the terms under which the LEP may deliver future secondary education facilities are also determined in the SPA.

BSF is a flexible programme: the length of the investment programme is not established *a priori*, but depends on the size of the local programme and the relative demands from other areas. The individual LEP's partnerships are also flexible; procurement depends on the willingness of all partners in the contract to continue and extend the co-operation. The partnership is also designed to be adapted to local circumstances and requirements.

Process

When the contracting authorities (LEAs) decide to start a BSF programme, they need to select a private sector partner for participation and investment in the LEP, to be established jointly with some or all of the contracting authorities. The selection process is comparable with the concession procurement method used in the UK. When the private sector partner has been selected in accordance with EU procurement rules, the LEP for a BSF area is set up and outlines the terms under which it will operate, including adhering to a business plan approved by all three shareholders.

The LEP will recover its costs and earn returns through successful delivery and operation of the contracts it manages. In addition to the possibilities offered by the initial contract, the LEP will be incentivized to develop and initiate proposals for other investments in secondary education which meet the requirements of the local authority. The LEP is also responsible for the provision or arrangement of partnering services, which implies services related to the development and implementation of a strategic investment programme and related to programme management and development. The partnering services need to comply with a two-stage approval process for each project and demonstrate continuous improvement and value for money.

The facilities to be provided will include not only new buildings and their services, but will also be procured through refurbishment of existing properties. Projects may include hard and soft facilities services; they may be provided through separate special purpose vehicles (SPVs) or companies (SPCs) established by the LEP, with which some of the contracting authorities may deal. The LEP is not obliged to take commercial risks to deliver approved projects through its supply chain. Probably, in most cases the LEP will contract with the local authority to deliver approved projects either directly or through SPVs or SPCs, and will therefore bear a number of the commercial risks in relation to these contracts. However, there is also a possibility for the private sector partner itself to take the commercial risks of delivering approved schemes, rather than the LEP. In such cases, the contractual arrangements will need to be approved by all three LEP shareholders. To bear and manage the involved risks, it is necessary for the LEP to be adequately capitalized. The capital required needs to be provided either by its shareholders as equity (in most cases) or by banks or other lenders as debt.

The long term character of the programme is expected to create scope for an increase in provider capacity and expertise to be achieved over time. The programme also generates packages and incentives for private actors to develop additional capacity through investments in innovative solutions that meet the exacting design and quality standards required, but would not be profitable to develop in the absence of a strategic investment programme.

The larger scale of projects compared to that of concession arrangements will be more attractive to potential private sector partners. The transaction costs of the procurement process can be spread over several properties and facilities, which will result in proportionately lower bid costs for both the public and private sectors. Furthermore, the knowledge gained in delivering the initial project provides opportunities for efficiencies and raised standards in the delivery of other education facilities.

Preliminary results

The first wave of projects in the BSF programme, announced in February 2004, involved so-called pathfinder projects and included five secondary schools, two special educational needs schools and an inclusion centre and support college. It comprises a mix of new build, remodelling and refurbishment to be delivered through a mix of concession arrangements and traditional funding. Through 2005, most authorities involved in this first wave developed their funding, educational visions and detailed physical school plans with their public partners; the first projects are now ready to start the EU procurement processes.

A point to be considered in BSF is the potential problems it generates in financial terms. Financing will not be dependent on specific projects, which may cause problems with respect to the important scrutiny effect that projects will lose, which are usually provided by private financiers and rating agencies.

8.5 Enhancing PPPs

Today, the strategic discussion on PPP is still driven by political feelings; in most cases it is not a rational debate. This is not surprising since no single conclusive statement on the performance of PPP based on empirical data can be given. The debate, furthermore, is troubled by the confusion about the term PPP. A clear understanding of the term PPP together with the different types of PPP is the first step towards a more open debate on the role and future of PPP. This has been one of the intentions of this book.

We have analysed the various forms of PPPs as well as the rationale underlying the various types. As we have seen, the choice of a certain type of PPP is often not a strategic one, but rather is based on cultural-historical developments or even coincidences, such as which advisor or financial institution was involved in the first pilot project. However, the different types of PPPs are not clearly distinguished in the literature, despite their differences in background and rationale. The choice between a concession and joint venture should, however, be of strategic relevance.

Public discussion confuses the provision of public goods and the safeguarding of public interest. Public goods might be provided by the private sector, with the government remaining responsible for the quality and accessibility of these facilities and services. As shown in Chapters 6 and 7, the public sector is still able to maintain control over supply, quality and prices in joint-venture PPPs. Safeguarding public interests does not imply that the government should deliver the facilities and services itself, but it must retain control over these to a certain extent.

However, some of the current PPP arrangements have fallen short of expectations. This is what we expressed as the lack of product and process performance in Chapter 2, and illustrated by the examples in the following chapters. Improvements in product performance do not always coincide with enhancement of process performance, but both aspects should be taken into account. In this book, several solutions have been described for improving elements of performance. The examples described in this chapter contain interesting lessons. In the next section we will further elaborate on current trends.

8.5.1 New trends in PPPs

Three types of advancement in PPPs can be seen:

- From project to portfolio management
- From single-function to integrated project PPPs
- From project- to more policy-based partnerships.

The first of these involves a way in which to introduce economies of scale to deliver more value for money. The key motives for a portfolio approach are related to flexibility, opportunities for innovation and reduction of transaction costs. A portfolio approach may also lead to less competition or even a monopoly position for private sector companies in the construction sector, although current practices in Germany and the UK give no indication of this. As indicated before, the optimal PPP arrangement does not exist, one should always find a good balance between public interest, quality and cost efficiency.

The second change is related to a shift to more integrated approaches: since space is scarce, spatial functions should be integrated. Because of the requirement for optimal integration of coherent functions (multi-function approach), interdependent stakeholders need to be involved (multi-actor approach). For every single stakeholder an optimal equilibrium between added value and investment should be found. Co-operation is expected to produce added value for all actors. At present, investments are often made on the basis of specific and functional considerations instead of on possibilities for combining functions (Bult-Spiering et al., 2005). However, actors are mutually dependent, which entails certain consequences: an optimal solution will not be found without co-operation. Due to growing dynamics and complexity, managing integral projects will become the challenge for the future. Joint-venture PPPs are seen as solutions for handling the complexity in these projects.

The third advancement is connected with the trend towards more policy-based joint ventures. To be able to influence the design of the facilities to be provided, private partners insist on early involvement in the process. It is striking, however, that the private sector considers early involvement more important than their public sector counterparts (Smit & Dewulf, 2002; Ahadzi & Bowles, 2004), indicating that the government is still convinced that it is the sole actor responsible for defining what is best for society. The LIFT cases are clear examples of more policy-based partnerships. In these portfolio partnerships, both the public and private sector partners are responsible for strategic planning, and the partnerships are designed to be adaptable and flexible in terms of meeting evolving needs. This indicates that the private sector partner is becoming more involved in strategic issues and processes. This is also the case in some joint venture arrangements in the USA, as described in Chapter 7. Typical for these arrangements is that public and private partners are becoming mutually responsible for and mutually dependent upon long term common goals. As shown in Chapter 2, a key condition for the success of projects in modern networks is the mutual dependency of partners in achieving their goals. Sharing responsibilities, revenues and risks for multiple projects over the long term are the characteristics of policy-based joint ventures, which contrast with the more traditional transfer of risks and revenues in the project-based concession arrangements. In concession PPPs one could argue that the term concession is referring to its other meaning, namely settlement, because many concession contracts result in claims and compromises. In policy-based partnerships, the focus is more on co-operation and alignment of goals at a multiple project level.

8.5.2 PPP as a strategic choice

As stated several times before, there is no single best PPP approach: the choice for a certain PPP arrangement should, however, be a strategic one. Most important for the choice of a particular PPP arrangement is that the reasons for involving private sector actors in concession projects are merely financially driven, while for joint ventures both project-based and policy-based characteristics, quality and synergy are the central motives. To safeguard the public interest, a government should not contract out the management of inner-city projects, because the public interest is diverse and cannot be safeguarded by means of a standardized contract as is possible for infrastructure projects. In urban area development we therefore see joint venture partnerships, while concessions are the main type of partnerships in the infrastructure sector.

The strategic choice for a certain PPP arrangement should be determined by the answers to two basic questions (Smit & Dewulf, 2002). The first question is: do the intended facilities and services create an added value to the core activities of the government? The second question pertains to the extent of activities that belong to the core competence of the department. Based on the answers to these questions we can make a choice, as pictured in Fig. 8.3.

In this respect a comparison can be made with the strategy of outsourcing in business organizations. Horgen et al. (1999) argue that activities with a low added value to the organization and which do not belong to the core activities may be contracted out easily. Parallel to this, if the activity is not a core competence of the government, but has an important public added value, some kind of strategic

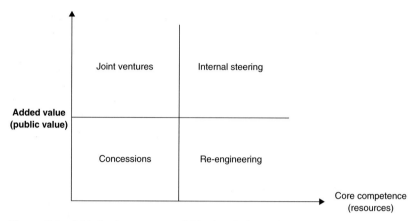

Figure 8.3   Added value to core activities in relation to the extent to which activities are core competences (Smit & Dewulf, 2002).

alliance with an outside provider is the most suitable solution to perform the activities. Nevertheless, as Savas (2000) indicated, changes in social value lead to changes in the way society determines what is a worthy collective good and what is not. In other words, the public interest axis is not static but dynamic, and is influenced by changes of environment.

Besides internal government considerations, possibilities in the marketplace should also be analysed. The duration and structure of the contract will depend both on the quality of the services provided by private parties and on safeguarding of the public interest. The public interest is not a stable concept but a dynamic one. As in the case of the LIFT structures, portfolio partnerships are aimed at generating more flexibility to meet local dynamic needs and are therefore a tool to deal with changing public interests.

8.5.3 Research agenda for the future

All chapters in this book have indicated that there are many gaps in our knowledge. Despite the numerous publications on concession PPPs, a conclusive answer concerning the performance of concessions cannot be given. Despite the high expectations of and wide public attention to joint ventures, little empirical evidence on their formation and functioning is presently available.

Integral studies are needed

Empirical research is needed to be able to have a more rational and fundamental discussion on the impact of and opportunities afforded by PPPs. Although every single evaluation study on a specific performance indicator could help to understand PPP arrangements, more attention should be given to a more integrated approach. Many interests are involved and a multi-criteria analysis is therefore needed. Moreover, fundamental changes are needed for the development and use of the public sector comparator. Open communication about the rationale of and interests in PPP projects is the first step towards improving project performance. Multi-disciplinary teams should work on studies related to PPP. Most studies on PPP address a single aspect of performance and study projects from a single perspective, e.g. financial, legal or organizational. As stressed many times before in this book, these aspects are intertwined and should be studied as a coherent entity. The performance of PPPs is not subject to one single condition but to a variety of interdependent variables.

Studies on new arrangements in PPP

The portfolio approaches described in this chapter will be the subject of further research. At Imperial College, London, the BEIC centre started several projects on new arrangements in healthcare. In our department at the University of Twente, Anneloes Blanken, co-author of this chapter, is doing a PhD study on portfolio approaches. Inge de Kort of our department has started a PhD study on management approaches for integral area development. In the years to come we can expect many interesting publications in this field.

Action-oriented research

Learning by doing is an oft-heard statement by people working in practice. However, it is the role of academics to generate insights, to build up theories and to develop innovative tools which can be used in that respect. Therefore, it is important to test and check academic knowledge and tools in practice. Academic and practical research are often considered to be at odds with one another; it should not be like that. Action-oriented research in which practice and science co-operate is an important condition for generating and disseminating new insights in PPP.

The role of process management

PPPs are inter-organizational relationships and take place in a network of interacting parties. Most researchers subscribe to these characteristics. However, only a little attention is paid to the problems related to the management of these networks. Most academic publications focus on the creation of a PPP, not on its functioning. An explanation for this may be the little attention paid by public agencies, industry and media to the process of functioning. When the PPP deal has been closed, it ends up in the archive and is no longer considered an interesting topic or object of study. In the opinion of the authors, however, managing successful PPPs is just the same as managing the interests of stakeholders.

The neglect of overall process management also explains why sociological aspects are ignored in the bulk of studies. Commitment, trust, perseverance and other social factors are, however, important conditions for success. Fortunately, we have noticed an increased interest in these intangible aspects in recent studies and national innovation programmes. In the Dutch national construction programme PSIBouw and in the Swedish programme Competitive Building, for instance, many projects have been started concerning the role of integrity, transparency and trust in innovative procurement (Atkin et al. 2003; www.psibouw.nl; www.competitivebuilding.org). In our department, as part of the PSIBouw programme, a large-scale research project started in 2004 on the role of trust in the functioning of partnerships for which a comprehensive survey has recently been undertaken; the results of this study will be published in 2006.

In accordance with the attention to sociological aspects is the development of research on the issue of communication in construction. Open communication and clear definitions of the requirements of the stakeholders are essential elements for the functioning of PPPs. In this respect, we can mention the work of Imperial College, London, on design quality indicators and the University of Salford's programme on process protocol. In a joint project with Imperial College, our department is further involved in the development of tools to improve communication in construction.

Related to all of these aspects is the impact of new innovative arrangements on public interest. Marnix Smit, co-author of Chapter 6, is doing a challenging PhD study on the relation between PPP and public interest. Several aspects of his study have been discussed in this book.

Professionalism

Innovative arrangements, such as portfolio partnerships and integrated approaches, test the skills and capabilities of both the public and private sector partners. In this book we have highlighted the importance of professionalism. Besides research in this field, more attention must be paid to education both in graduate programmes and in postgraduate courses.

8.6 Closing remarks

Many of the issues discussed in this book have been subject of debate for a long period of time. We realize that we have not provided clear-cut solutions or definite answers on the issues discussed. In consequence, critics might state that we have left the reader with many unanswered questions. However, we are convinced that a book on PPPs written in 2030 will still not have all the answers to these questions. This gap in knowledge may not, however, be an excuse not to think and act strategically.

The intention of this book was to provide materials and tools for the reader to make well-founded decisions on PPP projects. We have discussed the diversity in PPP schemes and elaborated on bottlenecks and the conditions for success. We hope that the reader has gained an improved understanding of issues related to the creation and functioning of PPPs, and that these insights will help them make strategic choices in the future.

References

Ahadzi, M. & Bowles, G. (2003) Public–private partnerships and contract negotiations. *Construction Management and Economics*, **22** (9), 967–978.

Atkin, B., Borgbrant, J. & Josephson, P. (2003) *Construction Process Improvement*. Oxford: Blackwell Publishing.

Audit Commission (2003) *PFI in Schools: Findings from Early Schemes*. London: Audit Commission.

Audit Scotland (2002) *Taking the Initiative, Using PFI Contracts to Renew Council Schools*. Edinburgh: Accounts Commission.

Ball, R., Heafy, M. & King, D. (2001) The private finance initiative – a good deal for the public purse or a drain for future generations? *Policy and Politics*, **29** (1), 95–108.

Boyne, G.A. (1998) Public services under new labour: back to bureaucracy? *Public Money and Management*, **18** (3), 43–50.

Bult-Spiering, M., Blanken, A. & Dewulf, G. (2005) *Handboek Publiek–Private Samenwerking*. Utrecht: Lemma.

Department for Education and Skills (DfES) (2003) *Building Schools for the Future: Consultation on a New Approach to Capital Investment*. Nottinghamshire: DfES Publications.

Edwards, P. & Shaoul, J. (2003) Partnerships: for better, for worse? *Accountability Journal*, **16** (3), 397–421.

Forsyth, T. (2004) Building deliberative public–private partnerships for waste management in Asia. Paper presented at *Conference on Democratic Network Governance*, 21–22 October 2004, Rosklide University, Denmark.

Glaister, S. (1999) Past abuses and future uses of private finance and public private partnerships in transport. *Public Money and Management*, **19** (3), 29–36.

Grimshaw, D., Vincent, S. & Willmott, H. (2002) Going privately: partnership and outsourcing in UK public services. *Public Administration*, **80** (3), 475–502.

HM Treasury (2003) *PFI: Meeting the Investment Challenge*. London: HM Treasury.

Institute for Public Policy Research (2001) *Building Better Partnerships: the Final Report of the Commission on Public Private Partnerships*. London: Biddles.

Horgen, T.H., Joroff, M.L., Porter, W.L. & Schön, D.A. (1999) *Excellence by Design*. New York: John Wiley & Sons.

Kirkpatrick, I. (1999) The worst of both worlds? Public services without markets or bureaucracy. *Public Money and Management*, **19** (4), 7–14.

McCabe, B., McKendrick, J. & Keenan, J. (2001) PFI in schools – pass or fail? *Journal of Finance and Management in Public Services*, **1** (1), 63–74.

Maltby, P. (2003) Comparing cost. *PFI Journal*, April 2003.

Mott MacDonald (2002) *Review of Large Public Procurement in the UK*. London: HM Treasury.

National Audit Office (1998) *The First Four Design, Build, Finance and Operate Road Contracts*. House of Commons Paper (HC) 476, Session 1997–98. London: National Audit Office.

National Audit Office (NAO) (1999a) *Examining the Value for Money of Deals under PFI.* Report of Comptroller and Auditor General, HC 739, Session 1998–99. London: The Stationery Office.

National Audit Office (1999b) *The PFI Contract for the New Dartford and Gravesham Hospital.* HC 423, Session 1998–99. London: National Audit Office.

National Audit Office (2005) *Innovation in the NHS: Local Improvement Finance Trusts.* Report by the Comptroller and Auditor General, HC 28 Session 2005–2006. London: National Audit Office.

Office of Government Commerce (2000) *Delivering a Step Change: OGC's Strategy to Improve Government's Commercial Performance.* London: Office of Government Commerce.

Partnerships for Schools (PfS) (2004). *Building Schools for the Future: the LEP Model,* vol. 1. London: Partnerships for Schools.

Rutherford, J. (2003) PFI: the only show in town. *Soundings,* **24** (1), 41–54.

Savas, E. (2000) *Privatization and Public–Private Partnerships.* New York: Chatham House.

Smit, M. & Dewulf, G. (2002) Public sector involvement: a comparison between the role of the government in private finance initiatives (PFI) and public private partnerships (PPP) in spatial development projects. In: *Public and Private Sector Partnerships: Exploring Co-operation* (pp. 451–463) Sheffield: Sheffield Hallam University.

Smit, M., Dewulf, G.P.M.R. & Bult-Spiering, W.D. (2002) *De Publieke Visie op PPS.* Enschede: P3BI.

Walsh, K. (1995) Competition for white collar services in local government. *Public Money and Management,* **5** (2), 11–18.

4Ps (2004) *A Map of the LIFT Process, Project and Procurement Support for Local Authorities.* London: 4Ps.

Further Reading

Dewulf, G., Bult-Spiering, M. & Blanken, A. (2004) *Opportunities for PFI in The Netherlands.* Enschede: P3BI.

Flinders, M. (2005) The politics of public–private partnerships. *British Journal of Politics and International Relations,* **7** (2), 215–239.

Heald, D. (2005) Value for money tests and accounting treatment in PFI schemes. *Accounting, Auditing and Accountability Journal,* **16** (3), 342–371.

Lonsdale, C. (2005) Post-contractual lock-in and the UK private finance initiative: the cases of national savings and investments and the Lord Chancellor's department. *Public Administration,* **83** (1), 67–88.

Partnerships for Schools (PfS) (2005) *BSF Standard Document.* London: Partnerships for Schools.

Salford Centre for Research and Innovation (2005) *Fuzzy front-end of design in the NHS MaST LIFT primary healthcare projects: research report.* Salford: University of Salford.

Index